Spacecraft Sensors

Spacecraft Sensors

Mohamed M. Abid
University of Southern California (USC), USA
University of California Los Angeles, Extension
(UCLA Ext.), USA
Jet Propulsion Laboratory (JPL), USA

John Wiley & Sons, Ltd

Copyright © 2005   John Wiley & Sons Ltd, The Atrium, Southern Gate, Chichester,
West Sussex PO19 8SQ, England

Telephone (+44) 1243 779777

Email (for orders and customer service enquiries): cs-books@wiley.co.uk
Visit our Home Page on www.wiley.com

Other Wiley Editorial Offices

John Wiley & Sons Inc., 111 River Street, Hoboken, NJ 07030, USA

Jossey-Bass, 989 Market Street, San Francisco, CA 94103-1741, USA

Wiley-VCH Verlag GmbH, Boschstr. 12, D-69469 Weinheim, Germany

John Wiley & Sons Australia Ltd, 42 McDougall Street, Milton, Queensland 4064, Australia

John Wiley & Sons (Asia) Pte Ltd, 2 Clementi Loop #02-01, Jin Xing Distripark, Singapore 129809

John Wiley & Sons Canada Ltd, 22 Worcester Road, Etobicoke, Ontario, Canada M9W 1L1

Wiley also publishes its books in a variety of electronic formats. Some content that appears in
print may not be available in electronic books.

Library of Congress Cataloging-in-Publication Data

Abid, Mohamed M.
   Spacecraft sensors / Mohamed M. Abid.
      p. cm.
   Includes bibliographical references and index.
   ISBN 0-470-86527-X (cloth : alk. paper)
      1. Space vehicles–Electronic equipment. 2. Detectors. 3. Astrionics. I. Title.
TL3000.A334 2005
629.47'2–dc22   2005005161

British Library Cataloguing in Publication Data

A catalogue record for this book is available from the British Library

ISBN-13 978-0-470-86529-9 (HB)
ISBN-10 0-470-86527-X (HB)

Typeset in Thomson Press (India) Limited, New Delhi

To my wife Michelle and my son Zander Ayman

Contents

Preface

As amazing as the human sensory organs are, they have natural limitations governed by our nervous system's ability to receive, interpret and analyze a particular phenomenon. Sensors in general are an extension to the human capability to obtain further knowledge of a given phenomenon. Optics allow us to see beyond the visible spectrum of light and to see objects at great distances. Antennas allow us to listen to deep space. Where our sense of smell leaves off, vapor chemistry takes over. Sensors use these technological advances to receive and analyze phenomena. Putting these sensors in space expands the scope of our knowledge of Earth, the atmosphere, other terrestrial bodies and on into space far beyond anything the human senses could ever attain alone.

Spacebased sensors is a field that is vast. There are entire books dedicated to each aspect of spacecraft sensors such as IR, remote sensing, signal filtering and radar, to name a few. A book that encompasses many of these fields together is rare, as it has proved difficult to cover so much information adequately in such a small amount of space. This book is an attempt to bring a basic under-standing of some of these components together to help understand how they work together in spacebased applications. I do not intend to discuss the results of given sensors, only to examine the basic understanding of the physics, how to formulate the phenomena, and derive from these equations a design that would allow us to perform the actual measurement. I intended to cover the physics, the math, the business, and the engineering aspects of sensors in one volume. The success of the development, construction and flight of a space-based sensor is entirely dependent on the team that is put together to develop each step. This team is usually composed of various dis-ciplines such as business, science, design and engineering. Often each discipline has its own language, philosophy and protocol. The success of each sensor, as my experience has proven, and the goal of this manuscript, is based on marrying these fields together and producing one comprehensive product.

Inspiration for this book came from my experience in industry and in teaching at the university level. For me teaching is something that I enjoy a great deal. Every course is always a learning experience, giving me knowledge that assists me in various undertakings in the professional world. After teaching some of the classic courses – such as 'Spacecraft Dynamics', 'Spacecraft Structure', 'Spacecraft Attitude Dynamics and Control' – the need to have a specific course and a book on spacecraft sensors seemed to become more and more necessary for the academic world as well as for the spacecraft industry. It has proved more than a challenge.

This book is a comprehensive review of spacecraft sensors for the specialist or non-specialist engineer, scientist and mathematician. It will cover sensor development from concept (physics, phenomenology), to design, build, test, interfacing, integration and on-orbit operation, as discussed in Chapter 1. Statistical analysis, error measurement assessment, noise reduction and filter optimization used in space operation will be discussed in Chapters 2 and 3. Typical sensors used in the spacecraft industry that will be covered are IR sensors, introduced in Chapter 4, passive microwave sensors, introduced in Chapter 5, active microwave or radars (Chapter 6) and spacebased GPS sensors, covered in Chapter 7.

This manuscript includes many examples from the space industry of past and currently orbiting sensors. These examples are set at the end of each chapter where the reader should be able to relate the various characteristics illustrated in the chapter to the working examples. The codes that generated some of the example plots, placed as a footnote, were done using Matlab. The purpose of these codes is by no means representative of code writing. It is only done for the purpose of giving the reader a starting point to extend these plots to cases that meet the needs of the reader.

This work would not be possible without the patient and moral support of my wife. I am very grateful as well for her diligent editorial work on the manuscript. Many thanks to my colleague, Jim Simpson, for reviewing the manuscript, and for his comments and discussions that helped me to steer the topics academically and professionally. Thanks to Brian Freeman for agreeing to be a coauthor, but because of his deployment he was not able to start his contribution. Many thanks to Gregor Waldherr, Paul Siqueira, Bruce Krohn and George Wolfe for reviewing various sections of the manuscript. Many thanks to the ULCA extension faculty for allowing me to introduce a course that helped me layout this book and to the UCLA Ext. students who helped me steer the direction of the course, which, in turn, guided the path towards the final manuscript. Many thanks to USC students who helped me to identify the area of interest that students want to study, understand and approach. Special thanks to my son for his beautiful smiles and giggles that helped keep me going, to my in-laws for their help and to my parents and brothers for their never-wavering support and laughter.

Signing off.

Mohamed Abid
Los Angeles, CA, USA

1

Introduction

Spacebased sensors are instruments that range in complexity from a simple thermocouple to a complex radar system. Regardless of the complexity of a sensor, there are a series of steps that are taken from the initial concept through design, budget proposal, development, construction, testing and launch to ensure a successful mission. Although there is no specific chronological sequence that must be followed, a general guideline is usually set in the early stages of development. This guideline is modified and enforced depending on the nature of the sensor, the mission and the funding agency.

A typical project life cycle is presented in Figure 1.1. Each agency or industry will have some differences in the execution, order and details from project to project. Critical steps are predefined between contractor and contractee, and an approval process is necessary in order to advance through each stage.

In the following sections, a general overview is presented of the various steps that a sensor may take from concept to on-orbit operation. This will include costs, trade-offs, environment, standard, packaging, interfacing, integration and testing.

1.1 Concepts

The development of sensors resides in the desire for the detection and measurement of a specific phenomenon for such purposes as

Spacecraft Sensors. Mohamed M. Abid
© 2005 John Wiley & Sons, Ltd

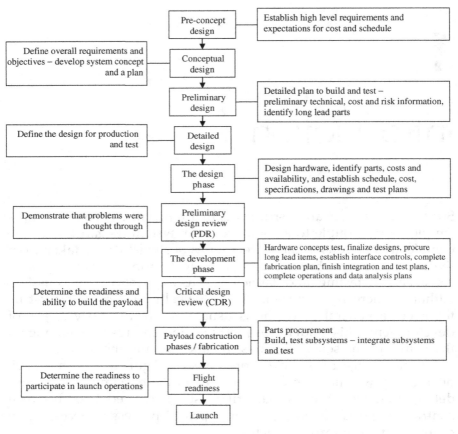

Figure 1.1 Typical project life cycle[2, 19, 21]

surveillance, weather pattern and forecasting, or for monitoring the health, attitude and orbit of a spacecraft. There are direct and indirect methods to conduct a measurement or an observation. For example, to measure temperature, a direct measurement can be taken using thermocouple technology. This is a quantitative, direct measurement taken without using an intermediate. Yet temperature can also be measured using touch. This is a qualitative and indirect measurement, since skin contact is used as an intermediate to detect heat or cold. The approach used to take a measurement depends on the maturity of the technology available and the cleverness of the developer.

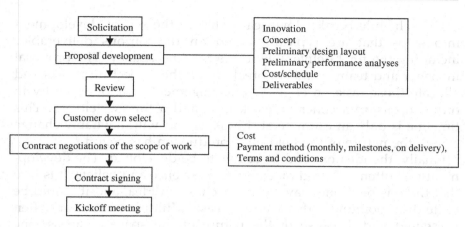

Figure 1.2 Typical proposal cycle

Sensor development can be initiated by many sources. It can begin with an inspiration to investigate or measure a phenomenon, then relate and justify the importance of that phenomenon and its impact on a mission or other project. The process can also start with a request for proposal (RFP) for a sensor or a project, where the funding agency such as the department of defense (DoD), NASA or the armed forces would publicly present a proposal guideline, a draft statement of work and a draft of specification. The RFP is then reviewed and evaluated by the funding agency. Figure 1.2 illustrates a typical proposal cycle. The proposal should meet the requirements of performance, schedule, cost or estimate. If all are within the specified goals, the process of formulating a proposal begins.

Although every proposal will be driven by the specifics of each project, the typical structure of a proposal begins first with cost, then schedule and finally concept as the primary elements. The budget weighs heavily in the bid process due to heavy competition between companies, agencies and contractors. There are two types of budget proposals: 'fixed' and 'cost-plus'. A 'fixed' budget is a proposed cost which maintains a budget cap, or limit. The 'cost-plus' budget means the project has an estimated cost, but the final awarded budget will reflect actual costs plus a handling fee, which may be a percentage of the final budget. The 'fixed' budget is the favored approach when there is a great deal of competition involved. The allocation of a budget is usually set and triggered to follow a fiscal year.

The schedule breaks down each step of the sensor development into steps that are typically based on the various deliverables including the final deliverable to the funding agency. The proposal manager and team members break down the project into tasks and allocate those tasks to engineers, management or another entity in order to ensure a coherent work flow and a timely delivery. This timeline is only an estimate at this stage and will inevitably change due to the nature of the field once production has begun.

Finally, the idea or concept begins to be developed. The development will often be based on experience, science and beliefs. It is rare that there is only one way of approaching an idea, and it would be quite disappointing if that were the case. Multiple options are often presented and, because of the nature of the space business, one would be chosen. For higher budget situations, or for low cost cases, more than one would be chosen, and these would later be narrowed down to one option based on the development process.

The commitment of funding for long-term projects is often enhanced when a proof of concept can be established and presented in the proposal. Proof of concept is an observation of the phenomenon in a simple form which suggests that the same phenomenon can be measured by a sensor in a more complex project. Although the proof of concept varies from project to project, the nature of the sensor, the extent of the project, the experience of the contractor and the requirement of the proposing agency make this step very crucial. For example, a project is proposed for a sensor based on the Sagnac effect. To establish a proof of concept, that phenomenon should be tested in every simplistic configuration. If the phenomenon can be observed simplistically, there is a greater chance that the more complex sensor project will succeed. It is common that the proof of concept is done by modeling or numerical simulation. This allows a better understanding of the parameters involved in the design phase.

Mission objectives and requirements are typically defined in a contract and/or statement of work, and the technical requirements are defined in the specifications. Contractual documents typically include at least three items: the contract, the statement of work and the specifications. The contract includes cost, payment plan, terms and conditions, proprietary data and hardware rights, in addition to any other legal issues. The statement of work lists the tasks to be performed, the contract deliverable requirements list, the schedule and program milestones and reviews. The specification includes the design requirements, the physical interfaces, the performance requirements, parts, materials and process requirements, and testing requirements.

1.2 Spacecraft Sensors Cost

This section explores the various approaches in cost estimation for spacecraft sensors that extend to all systems and subsystems. It will focus on sensors that constitute the main focus of the mission rather than sensors which simply assist in the general functioning of the spacecraft. For instance, when one considers the cost of developing and producing an attitude sensor in which more than one unit is built and is flown on many missions, the cost of this sensor is preset and the more this sensor flies, the lower its cost becomes. This type of sensor becomes an integral part of the overall bus function, such as solar panels. However, when a sensor constitutes the payload, such as a radar system, then the mission is designed around that sensor. Cost will then require a different approach for consideration due to the unique nature of that sensor.

1.2.1 Introduction to Cost Estimating

The purpose of cost estimation is to predict future costs for strategic decision making, market positioning, value chain analysis and life-cycle costing. It is difficult to estimate costs without detailed design because of the nature and the dynamic of the design changes. The need for accurate cost estimates was the impetus for the development of cost models. Cost models are highly valuable in predicting final costs. They evaluate cost trades between alternative concepts, support the selection of new concepts, evaluate cost sensitivities to technology selections, develop statistical cost metrics, provide preliminary budgetary information and provide data to decision makers.

The general approach is to define a cost object, known as the dependent variable, such as weight. Then determine the cost drivers that are known as independent variables. Once established, prior data is collected and interpreted using, for example, graphical analysis. This allows a study of the data trend and shape using regression, fitting and other methods that lead to a formal estimation. In this way the model is assessed and validated. It is important to re-estimate on a regular basis as more information becomes available which keeps estimates current as well as increases the accuracy of the estimate. In other words, the overall procedure of cost estimate prediction is to develop models that are based on collected data, then interpret the model and use it to predict an estimate. The more data that is collected, the more accurate the estimate will be.

Each project, mission or sensor requires a unique approach to cost estimation that is exclusive to that specific flight. Technology has not yet reached the level where one spacecraft bus or one part will be identical for all systems. Each project has a unique set of requirements, challenges and obstacles which require a specialized estimate. Common trends and generalized cost estimation approaches can be followed, but the final cost estimate will be unique to each project. Therefore, each cost estimate has a risk associated with it. This risk could be estimated using a range or a probabilistic distribution. Many programs have been terminated because costs were underestimated, expanse increased and funding was reduced. To minimize budget overrun and other such problems, risk analysis should be conducted. This analysis will determine sufficient cost reserves to obtain an acceptable level of confidence in a given estimate.

There are several cost-related terms which need to be properly defined. Direct cost is a cost that is directly related to designing, testing, building and operating the system. Indirect costs, often considered as overheads, are required for the survival of a business, but are not directly associated with development or operations such as management, profit or non-operational facilities. Non-recurring costs are only incurred once in a program such as research, design, development, test and evaluation, facilities or tooling. Because each project has unique qualities, this cost will be incurred on each mission as opposed to recurring costs which occur again and again throughout the life of the program such as vehicle manufacturing, mission planning, pre-flight preparation and checkout, flight operations, post-flight inspection and refurbishment, range costs, consumables such as propellants, and training. Refurbishment, a major contributor to space-flight cost, is the cost associated with maintenance and upkeep on reusable vehicles between flights. This relates mainly to launch vehicles, or space shuttle-like spacecrafts.[4]

Most cost estimates represent total life cycle costs (LCC) which include all costs to develop, produce, operate, support and dispose of a new system. It is very important to look beyond the immediate cost of developing and producing a system and consider all costs of a system's life cycle because what may appear to be an expensive alternative among competing systems may be the least expensive to operate and support.

The learning curve (LC) is a gauge of the experience of the company that is contracted to build, design or operate a system or subsystem. It could reflect the reduction of the labor costs when producing similar items – as an item is produced a number of times, efficiency increases and less labor is required each time it is produced. LC is described by the percentage reduction experienced every time production is doubled.[33] The uniqueness of each sensor to be flown makes the LC less important, but nevertheless present to be reflected in cost estimation.

Normalization is the process of adjusting the data so that it can account for differences in inflation rates, direct/indirect costs, recurring and non-recurring costs, production rate changes, and breaks in production and learning curve effects. It also accounts for unforeseen anomalies such as strikes, major test failures or natural disasters causing data to fluctuate. Cost estimators minimize distortion caused by cost-level changes by converting costs to a constant year dollar.

Often missions are set for a given period of time and cost is established accordingly. However, missions frequently last longer than scheduled. Critical decisions are made to decide on the implication to pursue a mission or to de-orbit the spacecraft. Alternatives such as selling the craft to other agencies or companies are exercised.

1.2.2 Cost Data

Compiling a database of cost data is very useful in making cost comparisons over time and between projects. In order to do this, an investment in a database that captures prior costs and technical data for proper cost estimate development is necessary. Unfortunately, this process is costly, time consuming and usually not funded.

There are two types of data sources. The first is the primary data found at the original source such as contractor reports or actual program data. The second type, which is less reliable but could be the best alternative when time constraints or data availability limit primary data collection, is derived from the primary data of another similar system such as documented cost estimates, cost studies/ research, proposal data, etc.. This data does not reflect a very accurate cost since it is interpreted as well as being an extrapolated type of cost.

In order to ensure that the cost data collected is applicable to the estimate it is necessary to identify the limitations in the collected data which are imperative for capturing uncertainty. Appropriate adjustments need to be made to account for differences in the new system to be considered. The bidding process also needs to be examined, since underbidding could be a motivation of a contractor to win the contract. This applies, as well, to data collected from sensors for phenomena interpretation or measurement analysis.

1.2.3 Cost Estimating Methodologies

There are many challenges in cost estimate model development such as limited time to develop estimates and resources. Once data has been collected and normalized, there are several methodologies available for estimating costs such as expert opinion, analogy, engineering, actual and parametric estimation. The choice of methodology adopted for a specific system cost estimate is dependent on the type of system and data availability. It also depends on the phase of the program such as preliminary design, development, build, integrate, fly and support development, production, or support.

The expert opinion method, as presented in Figure 1.3, is very useful in evaluating differences between past projects and new ones

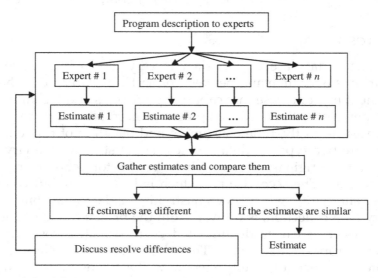

Figure 1.3 Methodologies: expert opinion – typical steps diagram

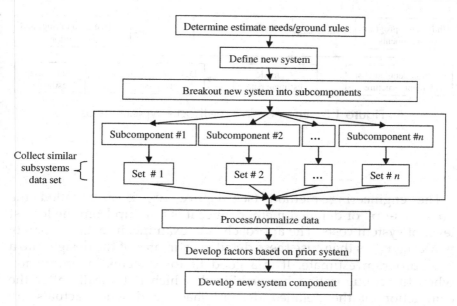

Figure 1.4 Method of analogy – typical steps

for which no prior data is available. This approach is suitable for new or unique projects. The drawback of this method is that it is a subjective approach because of the bias that could be introduced by an expert. Having a large number of qualified experts helps to reduce bias. The qualification of an expert is subject to interpretation.

The cost estimates using the analogy method as presented in Figure 1.4, rely on the comparison between the proposed programs with similar ones that were previously achieved. The actual cost is scaled for complexity, technical or physical differences to derive new cost estimates. This method is often used early in a program cycle when the actual cost data used for a detailed approach is insufficient. This method is inexpensive, easily changed and is based on actual experience of the analogous system. It is well suited, for example, for a new system is derived from an existing subsystem, a system where technology or programmatic assumptions have advanced beyond existing cost estimating relationships (CER). It also works as a secondary methodology or cross check. However, the drawback of this method is that it is subjective, and uncertainty is high since identical projects are hard to find to the extent that knowledge of the analogous programs mimic the cost of the new one.

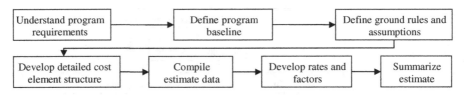

Figure 1.5 Engineering method – typical steps

The engineering method (see Figure 1.5) is also called the "bottoms-up" or detailed method since it is built up from the lowest level of system costs. The approach is to examine, in detail, separate work segments then fuse these detailed estimates of the design into a total program estimate. It is a good fit for systems in production when the configuration has stabilized, which is typically after the completion of the detailed design phase, and when actuals are available. This method is objective and offers reduced uncertainty but its process is rather expensive and time consuming.

The methodology actual bases future costs on recent costs of the same system and uses it later in development or production. It is a preferred method since costs are tuned to actual development or production productivity for a given organization. This method is the most accurate and most objective. Unfortunately, data is not always available in the early stages of the program.

1.2.4 The Cost Estimating Relationship Method

The cost estimating relationship (CER) method relies on statistical techniques to relate or estimate cost to one or more independent variables such as weight and software lines. For that purpose, CERs use quantitative techniques to develop a mathematical relationship between an independent variable and a specific cost. This method can be used prior to development and is typically employed at a higher cost estimate system level as details are not known. Most cases will require in-house development of CER. Figure 1.6 shows typical steps used for CER. The advantage of this method is in its forecast when implemented properly. It is also fast and simple to use, once created. However, there is often a lack of data for statistically significant CER, and it does not provide access to subtle changes.

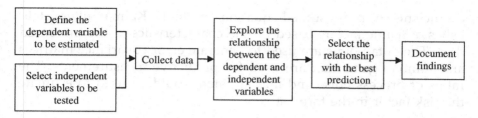

Figure 1.6 Typical steps for CER

A typical statistical technique that can be used to quantify the quality of the CER is the least mean squares (LMS) best fit, which is a regression analysis that establishes the ability to predict one variable on the basis of the knowledge of another variable. This can be extended to multiple regressions where a change in the dependent variable is needed to explain the dependent variable pattern. Estimation often requires extrapolation beyond the linear range of the CER as shown in Figure 1.7. Extrapolating CERs is often the best methodology available when limited or few data points are available which is typical of the space industry. Validity of the estimation outside of the range is based on the experience of the estimator and acceptance of the methodology by peers. Typically, an extrapolation is acceptable up to 20% of the data limits, after which additional risk must be assumed. The CER itself only provides a point estimate around which risk is applied.[4]

Instead of LMS, other techniques could be used such as a Kalman filter (see section 3.9). Unlike regression methods that generate

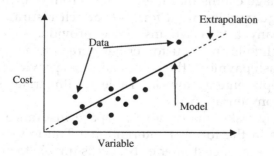

Figure 1.7 Data represented with a CER

coefficients for pre-assumed equation forms, the Kalman filter establishes dynamic factors based on the characteristics of the measured data. The data points are used to generate error covariance matrices and Kalman gains that are used for predictive measurements. The ratios of process noise and measurement could be used to estimate the risk factor in the forecast.

1.2.5 Insurance Cost

Insurance is becoming an intrinsic part of commercial missions. This could cover a sensor, an operation or the whole spacecraft. It is essential to understand the many different types of insurance, the volatile nature of the market, the level of necessity for each type of insurance and how events such as launch failure can affect costs.

There are many types of insurance, as one would suspect. The complexity of an orbiter and the amount of technology involved makes the decision difficult for commercial spacecraft producers to choose what to insure. For instance, pre-launch insurance covers damage to a satellite or launch vehicle during the construction, transportation and processing phases prior to launch. Launch insurance covers losses of a satellite occurring during the launch failures as well as the failure of a launch vehicle to place a satellite in the proper orbit. In-orbit policies insure a satellite for in-orbit technical problems and damages once a satellite has been placed by a launch vehicle in its proper orbit. Third-party liability and government-property insurance protect launch service providers and their customers in the event of public injury or government property damage, respectively caused by launch or mission failure. In the US, the federal aviation administration regulations require that commercial launch licensees carry insurance to cover third-party and government-property damage claims that might result from launch activity.[10]

Re-launch guarantees are a form of launch insurance in which a launch company acts as an insurance provider to its customers. When a launch fails and a customer has agreed to accept a re-launch instead of a cash payment, the launch services provider re-launches a customer's replacement payload. Table 1.1 illustrates some failures that lead to compensation.[10]

In evaluating risks, many non-US space insurance underwriters face obstacles in the form of International Traffic in Arms Regulations (ITAR). When a customer or broker is unable to obtain a license from the State Department to share a launch vehicle or a satellite's

Table 1.1　Claims for some recent failures

Failure	Spacecraft	Claims in million
Arian 5G upper stage	Artemis	$150
	BSAT-2B	
Orbital Sciences Taurus 2110	Orbivew 4	$75
Solar array	PanAmSat	$253
	Arabsat	$173

technical details with non-US underwriters, international insurers are forced to either decline the risk or to offer policies based on vague technical information.

1.3 Spacecraft Sensors Trade-off

There are two types of sensors – health sensors and payload sensors. Health sensors ensure and monitor the spacecraft or payload functionality such as temperature sensors, strain gauges, gyros and accelerometers. Examples of payload sensors are radar imaging systems and IR cameras. For health sensors, the spacecraft would list the requirement as to what would need to be measured or controlled for an optimum operation, and sensor selection would depend on this. One would not fly a spacecraft solely to test a gyro or a star sensor. Payload sensors represent some of the reason for the mission to exist. These sensors would set the requirements such as pointing, power consumption and operation.

Figure 1.8 illustrates a typical design trade-off evolution within a program. In this trade-off, the initial variables set the boundary or

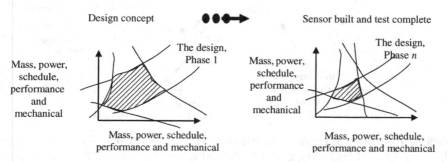

Figure 1.8　Typical trade-off study that leads to the ultimate design

constraint for the design, then as the development progresses, budget cuts, budget overruns, technology challenges and other changes narrow down options. Typically the hatched area in Figure 1.8 gets smaller and smaller to converge on the actual sensor design. There are two types of boundaries. The first type is hard boundaries that include high level requirements, mission constraints, space environment and physics law such as gravity. The second type is soft boundaries that are derived from high level requirements and flows all the way down to the part level. Soft boundaries are the ones that a sensor design would first consider to modify, adjust and trade with other boundaries. Typically soft boundaries can be changed at the system level. Hard boundaries can only be changed at the program and funding agency level.

1.4 Spacecraft Environment

The environment in space can either be a detriment to a mission, or it could be used to enhance mission performances. For instance, atmospheric density could be used to assist in re-entry, in removal of orbital debris and in aero braking. Plasma and the geomagnetic field could be used in spacecraft control, power generation and on-orbit propulsion such as electrodynamics tethers. The space environment, if not taken into account during the design process, can cause several

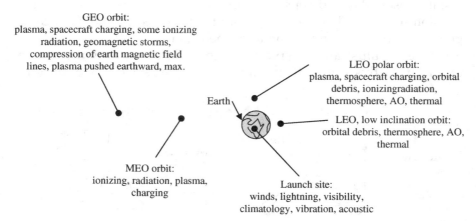

Figure 1.9 Space environments are orbit dependent (GEO = geostationary orbit, MEO = medium Earth orbit, LEO = low Earth orbit, AO = atomic oxygen)

spacecraft anomalies, from minor anomalies to temporary or permanent major damage. A spacecraft is subject to space vacuum, thermal cycling, plasma, atomic oxygen (AO), solar wind and many more phenomena. The level of importance of each environment depends on the orbit, as illustrated in Figure 1.9. The impact of this environment could lead to degradation and/or contamination, malfunction or permanent damage. Table 1.2 presents some of the environmental impacts on some typical surfaces.

The Sun, the largest source of electromagnetic waves in the solar system, has many activities that impact orbiting spacecraft in one way or another. Activities include solar wind, solar flares, coronal mass ejections and more. The Sun causes variations in the Earth's atmosphere from UV absorption (neutral and charged). Solar cycles with an 11 year period, are variations in solar output.

1.4.1 Vacuum

Without atmosphere, a spacecraft relies on transferring heat to its surroundings via thermal radiation or thermal conduction (see Chapter 4). Thermal control coatings reflect solar energy away and radiate thermal energy. Spacecraft materials may experience more than a hundred degrees of temperature change when going from sunlight to shadow. Degradation of these materials may have a significant effect on spacecraft thermal control.

Molecular and particulate contamination also contributes to degradation. For instance outgassing either causes molecules to be trapped inside the material or causes surface residues to leave the material. In order to control or minimize this contamination cold gas or chemical treatment before launch can be used. Contamination can also be minimized by using materials that outgas the least, by placing them some distance from sensitive surfaces such as optic surfaces, by using vacuum bakeout of the material before integration and by allocating time in the early stage of on-orbit operation for in-orbit bakeout.

1.4.2 Neutral Environment Effects

A neutral environment is created by two situations. The first is when mechanical effects such as drag, even with low pressure and physical sputter, create enough impact energy for rupturing the atomic bond. The second is when chemical effects resulting from atmospheric constituents create AO which can cause possible changes in surface

Table 1.2 Space environment effects[1]

	Neutral environment	Thermal environment	Plasma	Meteoroids / Space junk	Ionizing radiation
Definition	Atmospheric density variation, atmospheric composition, atomic oxygen	Solar radiation, albedo, radiative transfer	Ionospheric plasma, auroral plasma, magnetospheric plasma	Man made or other objects, size, mass, and velocity distribution	Trapped proton and electron, radiation, galactic cosmic rays, solar particle events
Electrical power	Degradation of solar array performance	Solar array designs, power allocations, power system performance	Shift in floating potential, current losses, contaminants trapping	Damage to solar cells	Decrease in solar cell output
Metals	Material degradation, oxides formation		Arcing, sputtering, contamination effects on surface properties	Material degradation	Material property degradation
Composite	Binder erosion. Fiberglass unaffected		Some charging, and arcing	Material degradation	Some material property degradation
Optics	SC glow, interference with sensors	influence optical design	Contaminants trapping, change in surface optical properties	degradation of surface optical properties	Darkening of windows and fiber optics
Thermal control	Reentry loads / heating, surface degradation due to atomic oxygen	Passive and active thermal control system design, radiator sizing, freezing points	Contaminants trapping, change in absorptance / emittance properties	Change in thermal / optical properties	
Mission operations	Reboost timelines, SC lifetime assessment	influence mission planning sequencing	Servicing EVA timelines	Crew survivability	Crew replacement timelines

Table 1.3 Atomic oxygen erosion rates: typical order of magnitude of annual surface erosion

Material	Silver	Mylar	Kapton	Epoxy	Carbon	Teflon	Aluminum
Erosion rate (mm/yr)	$\sim 10^{-1}$	$\sim 10^{-1}$	$\sim 10^{-1}$	$\sim 10^{-2}$	$\sim 10^{-2}$	$\sim 10^{-3}$	$\sim 10^{-5}$

condition through erosion/oxidation. Table 1.3 illustrates some of the erosion rates that depend on the material and the orbit of the spacecraft. The effects on the surface depend on the materials used but can include changes in thermal or optical properties.

Typical measures to take in order to minimize the effects of a neutral environment on the spacecraft are to further coat sensitive areas or to place sensitive areas in the lowest cross-sectional area of the spacecraft and orient it into the opposite direction of the spacecraft trajectory.

1.4.3 Plasma Environment Effects

Physical matter can exist in one of four states: solid, liquid, gas or a plasma state. Plasma is a gas which consists of electrically charged particles in which the potential energy of attraction between two particles is smaller than their kinetic energy. Spacecrafts can interact with both ambient and induced plasma environments. High-voltage solar arrays can be damaged by arcing. Floating potentials can charge a spacecraft solar array arc leading to damage on surfaces such as dielectric breakdown or sputtering due to ion impact. Currents collected by arrays flow in the spacecraft structure. Plasma oscillations occur when a few particles of plasma are displaced and the electrical force that they exert on the other particles will cause collective oscillatory motion of the entire plasma. If not properly considered, plasma could induce electrostatic charge/discharge of the spacecraft, dielectric breakdown or gaseous arc discharges. All of these are considerable problems for electronics and can cause physical damage of surfaces. Electrostatic discharge can be overcome by using uniform spacecraft ground, electromagnetic shielding and filtering on all electronic boxes. Electrostatic discharge tests verify the spacecraft's immunity to electrostatic buildup and its subsequent discharge to component and circuitry. A typical discharge from personnel during assembly or other electrostatic characterization of a human could be simulated by a 100 pF capacitor charged at about 3000 V and discharged through a 1.5 kΩ resistor to the case or the nearest conductive part of the sensor.

1.4.4 Radiation Environment Effects

Radiation in the environment can have varying effects on a spacecraft, its systems and any personnel within the craft. It can degrade solar cells, which reduce the power that they are supposed to generate. It can also degrade electronics, whose diffusion of charged carriers through semiconductor materials are relied upon by solar cells and other devices. Semiconductor's diffusion length (~1 μm) may be one order of magnitude smaller than for solar panels and therefore they are more sensitive. Radiation can also degrade the human cell reproduction process, which is a concern for manned flights of long duration. For this reason shielding is necessary. One of the many types of environmental radiation phenomena that can affect a mission is the Van Allen belts. These belts are composed of high-energy charged particles geomagnetically trapped in an orbit along earth magnetic field lines, which can penetrate the spacecraft and create ionization of the particles/atoms on their path. Other such energetic particles are associated with galactic cosmic rays and solar proton events. Nuclear detonations in space produce X-radiation, which results in high-temperature neutron radiation which results from fission, and gamma rays during burn. Delayed radiation in the form of gammas, neutrons, positrons and electrons from decay of radioactive fission products could also result. Electromagnetic pulse (EMP) is a secondary effect, due to X-rays and gamma rays ionizing the atmosphere and sending wave electrons. The impact of these particles on a system varies from one event to another. Possible scenarios involve either single event upset (SEU), where the sensor temporarily responds inconsistently with design characteristic, single event latchup (SEL), where the sensor does not respond to a signal, or single event burnout (SEB), where the sensor fails permanently.[17]

Particle radiation can displace and ionize materials in its path which results in the degradation of material properties. Electromagnetic radiation such as UV and soft X-rays causes degradation of material properties. Silica glass, thermal control coatings, some composites and ceramics may exhibit surface darkening upon exposure to this radiation. Simultaneous UV and contaminant flux on a surface can significantly enhance permanent contaminant deposition.

In order to minimize radiation effects, shielding structures should be placed between sensitive electronics and the environment in order to minimize dose and dose rate. Choosing an adequate material ensures a better response to radiation. This, however, would not apply to high-energy particles such as cosmic rays. For this case, redundant electronics are often used. Backup systems should be used in the event of a radiation problem, and a recovery procedure should be set up in the event of unforeseen radiation and its negative impact.

1.4.5 Contamination

There are many sources of contamination in space that result simply from the presence of the spacecraft in the space environment. For example, hydrazine thrusters firing would generate different molecules in size and composition, ion thrusters would generate particles and outgassing would generate volatiles. Volatiles would be, for example, deposited on the spacecraft and particles could impinge on surfaces. Both would primarily impact optic surfaces and thermal shields which affect the performances of these optic surfaces. Therefore contamination should be controlled by properly selecting the exposed material, and by positioning sensitive surfaces away from the source of contamination.

1.4.6 Synergistic Effects

The effect of the total space environment is often greater than the sum of the effects due to individual environments. For instance, the effect of atomic oxygen or contamination combined with the presence of UV would have a greater impact than if only atomic oxygen, contamination or UV were present individually.

1.4.7 Space Junk

Space junk (SJ) is generally composed of either naturally occurring particles or man-made particles. Naturally occurring particles include meteoroids and micrometeoroids whose average velocity is 17 km/s. Meteoroids include streams and sporadic natural particles that could be the cloud dust left over from a comet that might have been intercepted by earth. Man-made particles are called orbital debris

(OD) and have an average velocity of 8 km/s. OD can result, for example, from a spacecraft tank explosion. Debris that is larger than 10 cm could be tracked optically if the altitude is higher than 5000 km. Any altitude below that uses radar for tracking. For the most part, particles are smaller than 1 cm, which are not trackable, but shielding could be used to minimize their impact on the spacecraft such as pitting of optics, solar array and thermal shielding degradation. SJ threat is orbit dependent. For example, for low Earth orbit (LEO), high inclination orbits are the most susceptible to SJ especially between 700–1000 km and 1400–1500 km altitude. Geostationary orbit (GEO) orbits are less susceptible to SJ than LEO by a factor of 100. In order to minimize the damages, orbit selection, shielding using leading edge bumper and orientation are critical parameters to consider in the design stage.[31]

1.5 Standards

Standards present a specific approach and guideline for building the sensors or instruments. These guidelines apply from the component to the system level, through to integration and testing. Complying with these standards is crucial for a successful mission. Since a spacecraft is a whole congregation of systems and subsystems, which vary in functionality, performance, range and consumption, a complete coherence and a thorough understanding of the entire system enhances the success of the mission.

Each standard references other standards in one way or another. Some currently applicable standards date back as far as 1960 or earlier, although one may think that technology has evolved far beyond that early technology. This shows that a great deal of testing of a standard is necessary before it can be implemented. This can only bear witness to the high quality and fidelity of the supported technology.

NASA-STD or NASA-HDBK are referenced as guidelines, recommended approaches or engineering practices for NASA or other programs, both for manned and unmanned missions. Depending on requirements and constraints, these standards are referenced, in part or in full, in contracts and program documents.

The MIL-STD has broader uses. It ranges from the ground segment, sea, and air segment, to the space segment. In order to ensure a coherence, and interference-free mainframe, each segment should comply with the corresponding MIL-STD. For instance, in a complex environment where there is communication or interaction between these segments, such as in combat, it is very important to know that these entire segments can work coherently together.

In an attempt to consolidate references for MIL-STD, NASA-HDBK and other such references, a single clear structure is difficult to create, due to the broad range of standards documentation, the long history and the overlap of standards for all defense sectors. Nevertheless, a number of consolidated standards were created by the European space Agency (ESA) and have become the 'bible' of the European space standards. These approaches and goals were similar to the MIL-STD or NASA-STD and other government agency standards in the US. They provide a valuable framework for a wide spectrum of applications from systems with high integrity to less demanding applications. Figures 1.10 to 1.12 emphasize the structure of these standards in the following sectors – space engineering, space product assurance and space project management.

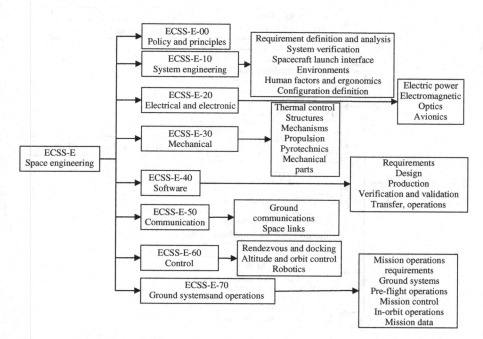

Figure 1.10 ECSS-E: European Cooperation for Space Standardization document tree for space engineering[7]

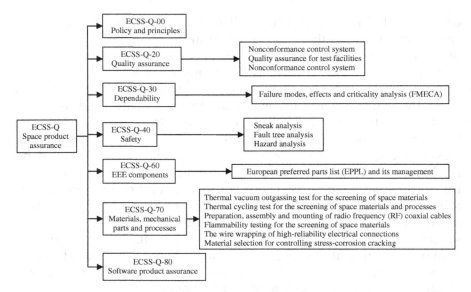

Figure 1.11 ECSS-Q: European Cooperation for Space Standardization document tree for space product assurance

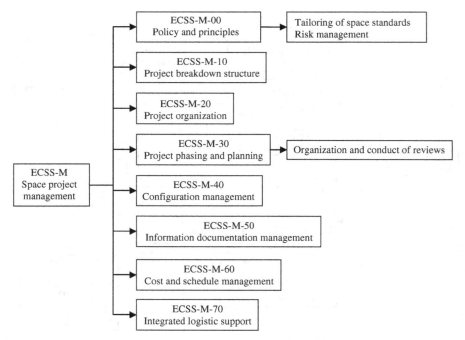

Figure 1.12 ECSS-M: European Cooperation for Space Standardization document tree for space project management

1.6 Packaging

The single common goal for spacebased sensors is how to 'fit a car in a match box'. This is a typical challenge in the space industry. Smaller means lighter, less power, more room for further instrumentation, and lower launch cost. The perpetual challenge is to achieve the same performance as a large scale instrument, if not better, and install it in a very small package. This field, even though it is not always identified as a specialty, is the essence of most design.

In the early history of spacecraft sensors, health monitoring sensors were mainly an extrapolation of existing sensors in the airplane industry. The primary difference resided in the packaging. Since weight, volume, power and shielding matter significantly in a spacecraft, extra care and engineering was performed to downsize and optimize existing technology to fit the needs of the spacecraft. Currently micro-electro mechanical systems (MEMS) technology is gaining ground in the space industry because of some of its attractive characteristics such as weight, size, robustness and reliability. Nevertheless, there is more work to be done for this technology to flourish like integrated circuits. MEMS are often compared to integrated circuits (IC) since they share many similarities – the fabrication process and silicon wafers are used for both MEMS and IC.

Packaging the sensor on the spacecraft is an art. Mechanically, structurally and electronically, the packaging has to comply with numerous requirements. Packaging may be required to protect the sensor from possible environmental damage for the duration of flight from takeoff all the way to on-orbit. Alternatively, it may be needed in order to fit in the launch vehicle. There are two types of packaging. The first type is fixed volume packaging. This type of packaging maintains the same form from integration all the way to the end of its life. The second type is the variable volume packaging. This type of packaging evolves from one step to another, such as deployment of radar antenna, and solar panels. Often the reason for this type of packaging is, for example, to fit the launch vehicle.

There are many optic packaging configurations where the reflector and the collimating optics are brought together to ensure that the incoming light is properly guided to the sensing element. The Newtonian configuration is a single large parabolic mirror which forms a real image at the axis of the telescope in the path of the incoming light (see Figure 1.13(a)). This configuration has a very limited viewing angle. The first stage of improvement is the

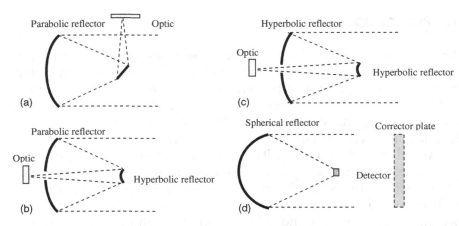

Figure 1.13 Typical configuration: (a) Newtonian, (b) Cassegrain, (c) Ritchey Chretien and (d) Schmidt[27]

Cassegrain configuration (Figure 1.13(b)). It retains the parabolic main mirror of the original Newtonian configuration but introduces a secondary mirror which may be made to a hyperbolic profile. The adjustable location of the second mirror can reduce the blockage substantially and therefore the system field of view is improved and the package size is reduced while maintaining a given focal length and performance characteristics. The Ritchey–Chretien configuration (Figure 1.13(c)) is a modified Cassegrain where the main mirror has a hyperbolic profile rather than the parabolic one. The use of two hyperbolic profiles provides a second additional degree of freedom in the design. This is the most used configuration in on-orbit telescopes. For wide-angle application, the Schmidt design is quite popular (Figure 1.13(d)). It uses a spherical mirror, in addition to a corrector plate that is mounted in front which provides two extra degrees of freedom and a second order like correction.

1.7 Interface and Integration

Typically the interface control document (ICD) defines the interface design requirements and the detailed interface design (see Figure 1.14). It provides the proper allocation of requirements across or at the interfaces. It is used to ensure that designs of the interfacing hardware, software or others are all consistent with each other and

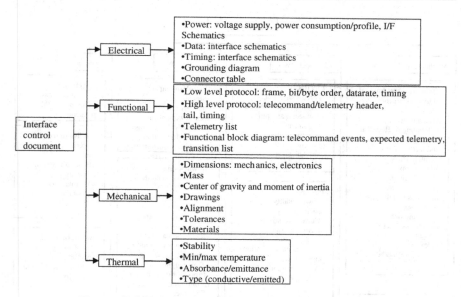

Figure 1.14   Interface control document (ICD)

that the requirements are verified. Figure 1.15 shows typical interface types that could be used.

1.7.1 Mil-STD 1553 Interface

One of many interfaces, as presented in Table 1.4, found in space-based systems is the 1553 interface which provides integrated, centralized system control for many subsystems connected to a bus. The 1553B, which is a variant of the 1553, provides flexibility and is a universally defined interface that allows compatibility between various end users. All subsystems are connected to a redundant twisted shielded pair of buses. Only one bus at a time is used. On the data bus, there are three functional modes of terminals. The first one is the bus controller (BC) which initiates information transfers on the data bus. It sends commands to the remote terminals which reply with a response. The bus supports multiple controllers, but only one controller may be active at a time. The second mode is the bus monitor (BM) often used for instrumentation. The 1553 uses the BM mode to extract information selectively for use at a later time. The third is the remote terminal (RT) which is used when neither the

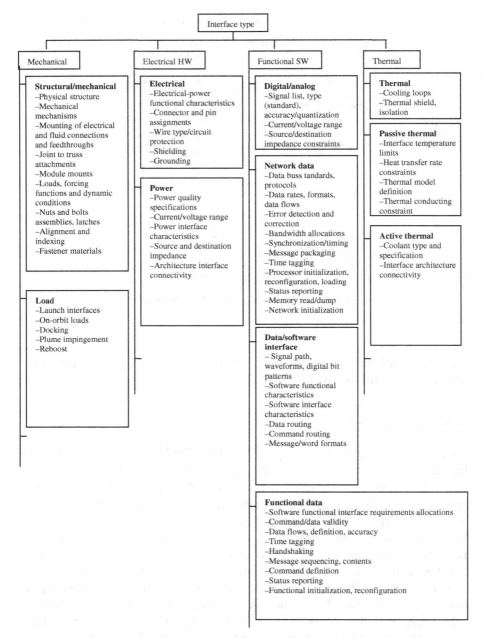

Figure 1.15 Typical interface types

Table 1.4 Typical interfaces, where MM and MS stands for master/master and master/slave

Interface	Configuration	Type	Timing	Control
RS-422	Star	Serial	Asynchronous/synchronous	MS/MM
RS-485	Bus	Serial	Asynchronous	MS/MM
PASTE	Star/Bus	Parallel	Synchronous	MS
CAN	Bus	Serial	Asynchronous	MM
Mil-STD 1553	Bus	Serial	Asynchronous	MM

BC nor the BM are used. Remote terminals are the largest group of bus components.

For the 1553 there are three types of words, the command word, the data word and the status word. The command word is transmitted by the BC and composed of three bits time sync pattern, five bits RT address field, one transmit/receive (T/R) field, five bits subaddress/mode field, five bits word count/mode code field and one parity check bit, for a total of 20 bits. In response to a BC request, the RT or the BC is used to transmit a packet with a 32 data word. A data word consists of three bits-time sync pattern, 16 bits data field and one parity check bit. A command message from the BC status words is transmitted by the RT. This status word consist of three bit time sync pattern, five bit address of the responding RT, 11 bit status field, and one parity check bit.

1.7.2 Proximity Issues

Placing or locating a sensor on the bus requires a complete analysis of the surrounding environment. For example, concerns for optics arise when a thruster is in the vicinity. The by-product of thrusters can harm the surface by either deposited by-products or by etching the surface of the optics. This is also true for thermal blanket and other sensitive surfaces. Testing involving electromagnetic interference and vibration, solve most proximity and compatibility issues. However, contaminations from outgassing or thrusters' by-products are less trivial to test when by-product trajectory is of concern. For an orbiter at high velocity, the issue is the dynamics of particles in a vacuum. Numerical simulation can address these issues and account for potential contaminants. The bake-out could give some information on the composition and the amount of the outgassed materials but it does not give any information on where and how the

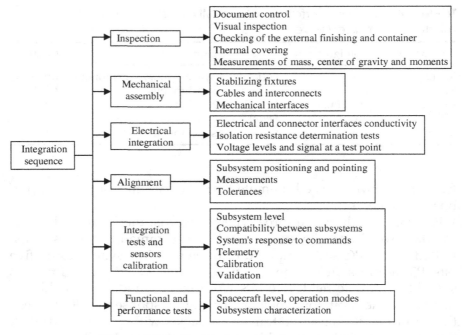

Figure 1.16 Typical assembly sequence

contaminant will interact with the surface upon impact. Modeling is often used to estimate the impact on outgassing or other particle contamination on the spacecraft.

1.7.3 Integration

A typical integration sequence is presented in Figure 1.16. In integration many of the interfaces stated in the previous section are verified. The level of scrutiny in the integration process is crucial. Along this process, electrical ground support equipment (EGSE) is used to validate integration and verify requirements. The order of the integration sequence, and the level of information needed in each sequence, is agency and mission dependent.

1.8 Testing

The goal of testing is to ensure and guarantee the functionality of each system and subsystem at all levels of integration and transition.

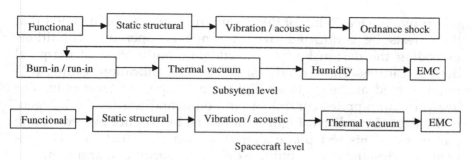

Figure 1.17 Typical environmental test sequence – the order may change from one company to another

Testing allows the verification that an instrument meets all performance requirements and that it is in compliance with requirements that are typically summarized in a system requirements verification plan of each mission. It must be verified that the instrument design will function in the spacecraft environment. A spacecraft sensor, like any other subsystem of a spacecraft, goes through a wide variety of environments where pressure, temperature, gravity levels, stress and other factors are constantly changing, until final orbit and end of life of the spacecraft. For these reasons, testing may include qualification vibration, thermal vacuum, radiation analysis and electromagnetic compatibility testing. A typical testing sequence is summarized in Figure 1.17. Component and payload level testing could differ. Identification and elimination of subtle design defects should be possible by setting up flight-like configurations and operating in different modes over extended duration.[23]

Before any testing, an approved test plan is often required. Test plans include requirements for the test, test approach, facilities, test flow, procedures, safety considerations, performance verification and validation. Typically, preliminary plans and procedures are delivered at the time of the critical design review (CDR). Changes and revisiting a prior test after already conducting one is performed when necessary, but this is often costly. Cost, schedule, planning, mission and agency, drive the testing sequence.

There are three test phases: development, qualification and acceptance. The objective of the development test phase is to ensure that testing of critical items at all levels of assembly are sufficient to validate the design approach. The development or prototype unit will prove the feasibility of the design. It is often recommended that these tests be conducted over a range of operating conditions exceeding the design limits, to facilitate identification of marginal design features. The second test phase is qualification testing. It is implemented at the flight or flight-like component level only. The test program applies controlled environmental test levels that generally exceed predicted flight parameters, but do not exceed design parameters. This test phase should demonstrate that each component, as designed and built, exhibits a positive margin on the performance specification. Table 1.5 lists the components that will undergo qualification tests for various types of hardware.

Numerical modeling or analysis is used for the optional tests shown in Table 1.5. An evaluation would determine whether the test has to be conducted or not, as well as the extent of the test. The third test phase is the acceptance test phase. This test should demonstrate that the assembled hardware will meet the performance specifications when subjected to controlled environments generally equal to the maximum expected flight levels. When a failure occurs, any corrective action could affect the validity of prior tests.

1.8.1 Performance Testing

The purpose of performance testing is to demonstrate that the hardware and software meet their performance requirements when provided with the appropriate stimulus. The first test conducted should serve as the baseline reference to determine if subsequent degradation of performance occurs. This test should be performed prior to and after each environmental test. Partial or limited performance tests should be conducted during transitions within the environmental test. For instance, performance tests should be performed between each axis of vibration.

1.8.2 Thermal Testing

The thermal environment of a spacecraft is harsh. The spacecraft can be exposed for long durations to Sun rays and extremely high temperatures, or it could be behind the Earth or another celestial body and blocked from the Sun producing extremely low temperatures. The impact of such extreme thermal environments can be dramatic for both flight hardware and sensors if these extremes were not anticipated during design phases. Hardware can exhibit unique thermal and mechanical properties under such stress. For

Table 1.5 Typical component qualification/acceptance tests: O stands for optional, R stands for required and both are mission and agency dependent; the last column is hardware integrated acceptance level test[23]

	Electronics	Antennas	Mechanism	Electro-mechanical	Deployable	Optics	Instrument	Structures	Flight hardware
Functional	R	R	R	R	R	R	R	R	R
Vibration	R	R	R	R	R	R	O	O	R
Acoustic	O	O	O	O	O	O	O	O	R
Ordnance shock	R	O	R	R	R	R	O	O	R
Mass properties	R	R	R	R	R	R	R	R	R
Static structural	O	O	O	O	O	O	O	R	O
Thermal vacuum	R	O	R	R	R	R	R	O	R
Thermal cycling	R	O	R	R	O	O	O	O	R
EMC	R	O	O	R	O	O	R	O	O
ESD	R	O	O	R	O	O	O	O	O
Corona	O	O	O	O	O	O	O	O	O
Magnetic properties	O	O	O	O	O	O	O	O	O
High/low voltage	R	O	R	R	O	R	O	O	R
Plasma	O	O	O	O	O	O	O	O	O

instance, materials can outgas under high vacuum conditions and destroy flight hardware. Oils and lubricants can overheat or freeze in space. Sensors can also be dramatically affected by extreme thermal environments if they are not anticipated and controlled. Outgassing in this situation can cause materials or species to land on an optic surface which could jeopardize the sensitive observation or measurements. A local freeze can cause the cracking or breaking of materials or can cause the seizing up of joints which ceases the mobility of the part. Extreme conditions can also cause materials to expand differentially and sensors that rely on alignment can go off alignment. Thermal swings can cause fatigue in a solder joint, for example, which can lead to shorting and potential circuit failure.

Thermal tests include thermal balance (to verify models), thermal cycling and thermal vacuum testing. The purpose of these tests is to demonstrate the survivability and performance under temperature conditions which exceed flight temperature levels. Temperature levels act as an environmental stress to stimulate latent defects and to prove the design. Emulation of the flight thermal conditions are manifested in test temperature levels which are based on worst-case high and low temperature extremes, with an added margin. These added margins are sufficient to act as a workmanship screen while also demonstrating thermal capability beyond expected flight temperatures. If flight component temperature variation is small, then component test temperature levels may be established to provide a minimum temperature differential between test levels to stress the component adequately. Test conditions and durations are dependent upon the test article's configuration, design and mission requirements.[23]

Thermal balance is used to verify the thermal model and thermal design validation. It is normally performed at the subsystem and payload levels and is used to verify the analytical thermal model and provide a confidence of that modeling. Thermal balance is typically conducted during thermal vacuum testing.

Thermal cycling is a test that is performed to detect design problems related to differential thermal expansion and defects related to latent material and workmanship. It is often completed prior to the installation of components for the next level of integration. Performance of the test at the component level allows necessary modification without impacting the whole system. It is also less expensive since a vacuum is not required. Thermal cycling is normally used to qualify space components.

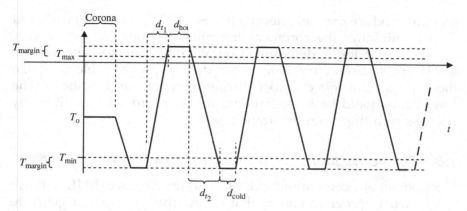

Figure 1.18 A typical thermal vacuum/atmospheric test cycle

Thermal vacuum testing is more expensive but is better in determining acceptability of thermal design. This test is performed to verify that the hardware can operate nominally in a vacuum environment at a wide range of temperatures which mimic hardware operation in space. This would validate the safe component operation before, during and after nominal mission operating temperature, at temperatures exceeding the extremes predicted for the mission and during temperature transition.

Testing should include the number of cycles used, the heating/cooling rate, location of thermal control, atmosphere versus vacuum or in the presence of a specific gas such as CO_2, and the duration of the test (see Figure 1.18). These parameters are agency dependent.

For the thermal vacuum test, temperature sensors used to monitor this test are often mounted on the component base plate or on the heat sink to which the component is mounted. An outgassing phase is included to allow volatiles to be removed from the test item. The outgassing phase is incorporated into a heat exposure that occurs during thermal vacuum testing. This is normally conducted in the first temperature increase where this high temperature is maintained until contamination levels are below a certain threshold.

1.8.3 Corona-arcing

Corona-arcing can be observed during pressure reduction. This could be a detriment to hardware. Low-power, low-voltage sensors will not produce corona. Thermal bake-out of sensitive items should help to minimize the corona-arcing phenomenon. The presence of corona and arcing is determined on all components where voltage levels are sufficient for it to occur during pressure bleed of the thermal vacuum test chamber during temperature transitions. The level item should be in operational mode in order to monitor any changes resulting from this transition.

1.8.4 Electromagnetic Compatibility and Interference Testing

Electromagnetic compatibility (EMC) and interference (EMI) tests are conducted in order to ensure that the hardware does not generate EMI that will adversely affect the safety, corruption and operation of the rest of the payload. This test also should ensure proper operation of the hardware in the presence of an external EM source. Testing allows the detection of unwanted emissions. The EMC/EMI testing is composed of four categories: conducted emission, conducted susceptibility, radiated emission and radiated susceptibility.

Conducted susceptibility tests check the instrument response to power spikes that could be generated from surrounding instruments, electrostatic discharge or other equipment coupling. This testing also verifies immunity to noise, ripple and transients on the interfacing power and signal lines. Conducted emission testing verifies the compatibility of components to be placed on the spacecraft. It is often measured in a shielded enclosure with flight-like configurations of cables and peripherals. These tests determine what the tested item generates and conducts outward on the interfacing power and signal lines. Radiated susceptibility tests are conducted on all instruments. These tests determine the level of susceptibility to specified EM environments. This would verify the immunity of the sensor against electromagnetic fields generated by other instruments. Radiated emission tests measure EM emanating from the tested unit and verify that the emanated EM are sufficiently small or do not impact other subsystems. This test would verify that spurs produced by the instrument will not interfere with the operation of other devices by exceeding the required level. Testing would measure the overall radio frequency (RF) output level of the tested instrument. Results are used against other equipment's susceptibility.

Figure 1.19 shows a typical setup where the instrument to be tested is located in an anechoic-like chamber. Anechoic chamber material

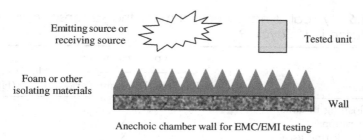

Anechoic chamber wall for EMC/EMI testing

Figure 1.19 Typical EMC/EMI testing setup

absorbs RF and generally minimizes interference between the internal and the external EM environment as well as reflection. This chamber also ensures that for a specific band most EM waves generated inside the chamber are attenuated in the walls. An emitting EM and/or a receiving EM are placed around or are scanned around the instrument at different distances.

1.8.5 Vibration Testing

During the life cycle of a sensor – from assembly, integration, liftoff to on-orbit operation and possible landing – each system and subsystem undergoes many dynamic environments (see Table 1.6). Vibration tests will allow the spacecraft to go through many simulated dynamic environments to determine whether the instrument will survive each step to orbit. Comparison of the actual structural behavior along with numerical modeling is often used to validate the test. Vibration testing determines the resistance of hardware to expected vibrational stresses. The sinusoidal vibration sweep determines the natural frequencies of the hardware. This survey also defines a baseline condition of the structure. The random vibration test may detect material, workmanship and design defects.[26] Typically, measurements are obtained using calibrated accelerometers.

Pyroshock tests are used to validate that shock does not affect components near the separation area. They are conducted on all components in the separation/deployment area (not limited to the launch separation system). For analysis a 3dB loss for each mechanical attachment is typically used. Sine sweeps and functional tests should be conducted prior to and post a pyroshock test to verify hardware condition.

Table 1.6 Typical dynamic environment that a sensor may experience during its life cycle[26]

Environment	Type of excitations			Max. frequency (Hz)
	Random	Transient	Periodic	
Liquid sloshing in tanks	X			5
Flight operations		X		10
Maneuvers during ascent		X		10
Seismic loads		X		20
Wind and turbulence	X			20
Liftoff release		X		20
Rocket motor ignition overpressure		X		40
Stage and fairing separations		X		50
Transportation	X	X	X	50
Engine/motor thrust transients		X		100
Pogo (interaction between liquid rocket engines and vehicle structure)			X	125
Solid motor pressure oscillation			X	1000
Engine/motor generated vibration	X		X	2000
Surface penetration		X		3000
Engine/motor generated acoustic noise	X			10 000
Aerodynamic sources	X			10 000
Onboard equipment operations	X	X	X	10 000
Planetary descent, entry and landing loads	X	X		10 000
Pyrotechnic events		X		100 000
Meteoroid impacts		X		—

1.8.6 Balancing

The knowledge of exact mass properties of the sensor system, subsystem and its location on the spacecraft bus is very important in order to identify the center of mass which is essential for attitude control, orbit operations and launch vehicle constraints. Parameters are determined for all configurations such as deployed or not deployed configuration. Hardware balance can be achieved analytically or by direct measurement. For a spin-stabilized spacecraft the impact of unbalance, such as an offset of the center of mass from the axis of rotation, would create precession and other dynamical movements that could jeopardize, for example, pointing requirements. Also crucial for attitude control is the knowledge of the moment of inertia, which is computed based on the mass distribution. This information is often computed as follows but not verified by measurement.

Typically, a sensor is a rigid body, which is a collection of particles constrained so that there is no relative motion of any one particle with respect to any other particles. The inertia matrix J that describes this sensor would be

$$J = \begin{pmatrix} I_{xx} & I_{xy} & I_{xz} \\ I_{xy} & I_{yy} & I_{yz} \\ I_{xz} & I_{yz} & I_{zz} \end{pmatrix} = \begin{pmatrix} \sum_i m_i (y_i^2 + z_i^2) & -\sum_i m_i x_i y_i & -\sum_i m_i x_i z_i \\ -\sum_i m_i x_i y_i & \sum_i m_i (x_i^2 + z_i^2) & -\sum_i m_i y_i z_i \\ -\sum_i m_i x_i z_i & -\sum_i m_i y_i z_i & \sum_i m_i (y_i^2 + x_i^2) \end{pmatrix},$$

$$(1.8.1)$$

where I_{xx} is the inertia about the system x axis, I_{yy} is the inertia about the system y axis, I_{zz} is the inertia about the system z axis, I_{xy} is the system xy product of inertia. Inertia matrix terms indicate the body's resistance to being angularly accelerated about different axes just as the mass indicates its resistance to linear acceleration. J is a symmetric matrix therefore it can be diagonalized by a coordinate rotation Q:

$$J_{P:\ \text{Principal axis}} = QJ_SQ^T = \begin{pmatrix} \lambda_1 & 0 & 0 \\ 0 & \lambda_2 & 0 \\ 0 & 0 & \lambda_3 \end{pmatrix} = \begin{pmatrix} A & 0 & 0 \\ 0 & B & 0 \\ 0 & 0 & C \end{pmatrix}, \quad (1.8.2)$$

where λ_i are the eigenvalues or principal values. Information regarding body symmetry is often helpful in locating the principal axis frame. If the body is symmetric about a particular plane then an axis normal to this plane is a principal axis. Using these symmetry properties for the case of a box with the dimension: a, b and c, the inertia matrix of each in their center of mass is

$$J_{\text{box}} = \begin{pmatrix} A & 0 & 0 \\ 0 & B & 0 \\ 0 & 0 & C \end{pmatrix}$$

$$(1.8.3)$$

$$A = m\frac{b^2 + c^2}{12}; B = m\frac{a^2 + c^2}{12}; C = m\frac{b^2 + a^2}{12}.$$

The inertia matrix at a point x_0, y_0 and z_0 different from the center of mass:

$$J_o = J_{CM} - m \begin{pmatrix} 0 & -z_0 & y_0 \\ z_0 & 0 & -x_0 \\ -y_0 & x_0 & 0 \end{pmatrix}^2, \qquad (1.8.4)$$

where J_{CM} is the inertia matrix in the center of mass. The total inertia of a system at a point will be the sum of all individual inertias of subsystems computed at this point of interest.

1.8.7 Mission Simulation Tests

The complexity of the spacecraft, the wide variety of fields that are necessary to operate the spacecraft and its payload teams make mission simulation critical to the success of a mission. This step provides the mission operators with an opportunity to operate the sensors using the in-flight operators, command procedure and telemetry database in launch configuration, early operation and normal operation. It also allows a look at mission modes and exercises how to enter into each mode of operation.

The testing process is overseen by a simulation team which runs the normal scenario of the mission and implants anomalies or malfunctions in some of the operational steps, to see how the operating team would recover from the anomaly by using alternative procedures. This is also a test of the operation procedures since they may include some steps that were overseen. Simulation often focuses on transitions, since problems mainly reside in transitions. Therefore operation in launch, separation, deployment and on-orbit operation should all be tested to examine each transition.

The number of these simulation tests is determined by the budget, and often a budget overrun and restricted time schedule may shrink the extent of these simulations and limit the number of times a test is run.

For manned missions, simulations are critical since crew interactions are often necessary. Therefore a coherent procedure list, as well as a well-operated communication link between crew and ground, become a natural obligation for multiple thorough simulations in order to ensure a successful mission.

1.9 Sensors on Orbit

Regardless whether a sensor is used for reconnaissance, attitude determination or remote sensing, a reference frame in which the sensor conducts the measurement needs to be well defined. The reference frame is determined in respect to a local reference frame or inertial reference frame for reconstruction of the measured information. In this section reference frames, coordinate transformation, orbit and attitude determination are introduced.

1.9.1 Reference Frame

To define the position of a point P a reference frame must be chosen and, for that to happen, origin and orientation need to be defined. Popular frames used are Cartesian, cylindrical and spherical, as illustrated in Figure 1.20. In order to obtain the coordinate of P an orthogonal projection of the point in all the axes must be conducted:

$$OP = x\boldsymbol{i} + y\boldsymbol{j} + z\boldsymbol{k} = \begin{pmatrix} x \\ y \\ z \end{pmatrix}_C \qquad (1.9.1)$$

where $\boldsymbol{i}, \boldsymbol{j},$ and $\boldsymbol{k}$ are unit vectors and O is the center of the reference frame. Any algebra used to manipulate physical quantities has to be conducted in the same reference frame. If nothing is specified, the same frame is assumed.

The Cartesian coordinate system is the most common and the easiest one to visualize; however, it is often not the easiest to use. For example, for a radial force applied to a sensor orbiting Earth,

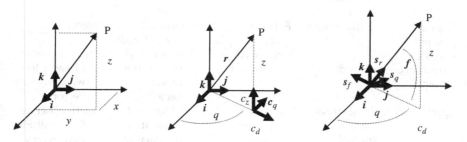

Figure 1.20 Cartesian, cylindrical and spherical coordinates

cylindrical or spherical coordinates would be more suited and easier to use.

$$r = dc_d + 0c_\theta + zc_3 = \begin{pmatrix} d \\ 0 \\ z \end{pmatrix}_{cy}$$

$$\begin{cases} c_d = \cos(\theta)i + \sin(\theta)j \\ c_\theta = -\sin(\theta)i + \cos(\theta)j . \\ \qquad c_z \end{cases}$$

(1.9.2)

Instead of using the coordinates (x, y, z) (three Cartesian coordinates), (d, z) are used (two cylindrical coordinates). The angle θ provides the azimuth angle of the unit vector c_d relative to i.

$$\text{OP} = rs_r = \begin{pmatrix} r \\ 0 \\ 0 \end{pmatrix}_s$$

$$\begin{cases} s_r = \cos(\phi)\cos(\theta)i + \cos(\phi)\sin(\theta)\sin(\theta)j + \sin(\phi)k \\ s_\theta = -\sin(\theta)i + \cos(\theta)j \\ s_\phi = -\sin(\phi)\cos(\theta)i - \sin(\phi)\sin(\theta)j + \cos(\phi)k \end{cases}$$

(1.9.3)

The number of variables is reduced from two variables (d, z) to one spherical coordinate, r. Table 1.7 illustrates some classic reference

Table 1.7 Common coordinate systems used in space applications[18]

Coordinate name	Celestial (inertial)	Earth fixed	Spacecraft fixed	Roll, pitch, yaw (RPY)
Fixed with respect	Inertial Space	Earth	Spacecraft	Orbit
Center	Earth, GCI: Geocentric inertial	Earth	Defined by engineering drawing	Spacecraft
Z- axis or pole	Celestial pole	Earth pole = celestial pole	Spacecraft axis toward nadir	Nadir
X-axis or reference point	Vernal equinox	Greenwich meridian	Spacecraft axis in direction of velocity vector	Perpendicular to nadir toward velocity vector
Application	Orbit analysis, astronomy, inertial motion	Geolocation apparent satellite motion	Position and orientation of spacecraft instruments	Earth observation, attitude maneuvers

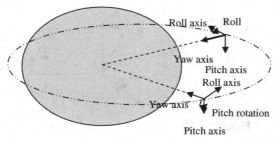

Roll, pitch and yaw coordinates

Figure 1.21 Roll, pitch and yaw coordinates[18]

frame or coordinate systems used in space applications. Figure 1.21 illustrates the RPY reference frame. It is not unlikely that for a mission, the reference frame changes during the mission. For instance, interplanetary missions start with Earth as the center or origin of frame and can then switch to a sun centered reference frame, and so on.

1.9.2 Coordinate Transfer in Three-dimensional Space

Within a mission, the reference frame used for a sensor could be very different, for example, from the one used by the attitude control system; but overall, any measurement conducted should easily be transformed from one reference frame to the other. In this section some typical transformations are introduced. For instance, let $I = XYZ$ be the inertial frame, and $S = xyz$ be the body fixed frame. The two frames I and S are related by a three by three transformation Q that changes a frame I to S such that

$$(a)_S = Q(a)_I$$

$$(a)_I = \begin{pmatrix} a_X \\ a_Y \\ a_Z \end{pmatrix}$$

$$(a)_S = \begin{pmatrix} a_x \\ a_y \\ a_z \end{pmatrix} \quad\quad (1.9.4)$$

$$\begin{pmatrix} x \\ y \\ z \end{pmatrix} = Q \begin{pmatrix} X \\ Y \\ Z \end{pmatrix}.$$

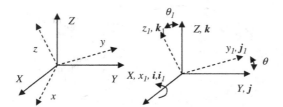

Figure 1.22 Frame *I* and frame *S*

The magnitude of a vector does not change with the frame regardless of what frame is used. Some of the properties of the transformation matrix Q are

$$Q^{T}Q = I \Rightarrow Q^{T} = Q^{-1}.$$
$$|Q| = \pm 1 \tag{1.9.5}$$

Such a matrix is called an orthogonal matrix. All transformations corresponding to a rotation of the coordinate frame are of this type. Using the determinant properties, if the determinant of Q is -1, this corresponds to a rotation from a left-handed to a right-handed frame. For instance, for the rotation Q of an angle θ_1 about the x axis of $I = XYZ$, $S_1 = x_1y_1z_1$ fixed in a rigid body, as illustrated in Figure 1.22,

$$Q \equiv Q_x(\theta)$$
$$XYZ \xrightarrow[\text{Around the axis: } X, x_1]{\text{Magnitude: } \theta_1} x_1y_1z_1. \tag{1.9.6}$$

The transformation of any vector from I to S is

$$\begin{pmatrix} a_{x1} \\ a_{y1} \\ a_{z1} \end{pmatrix}_{S_1} = Q_x \begin{pmatrix} a_X \\ a_Y \\ a_Z \end{pmatrix}_I \Rightarrow \begin{cases} (a)_{S_1} = Q_x(a)_I \\ Q_x = \begin{pmatrix} 1 & 0 & 0 \\ 0 & c\theta_1 & s\theta_1 \\ 0 & -s\theta_1 & c\theta_1 \end{pmatrix} \end{cases} \tag{1.9.7}$$

Q_x indicates that the rotation was about x. For sensors transformation, let I and S be related through Q

$$G_S = QG_IQ^{T} \tag{1.9.8}$$

This is very useful whenever, for example, the inertia matrix for a rigid body is used:

$$J_S = Q_y(\theta)J_I Q_y^T(\theta)$$
$$J_I = Q_y^T(\theta)J_S Q_y(\theta)$$

(1.9.9)

1.9.3 Conic Trajectories

Using Kepler's and Newton's laws one can derive the orbit of the spacecraft. The orbit of a spacecraft under a central force is described by

$$r = \frac{p}{1 - e \, \cos(\theta - \theta_0)},$$

(1.9.10)

where p is the parameter, e is the eccentricity and θ is the true anomaly. This equation describes a conic, which is the intersection of a plane with a cone (see Figure 1.23). A conic could be a circle, an ellipse, a hyperbola or a parabola. Table 1.8 shows the various properties of conics.

Figure 1.24 shows elliptical orbit geometry and the parameters that describe it. Figure 1.25 shows hyperbolic orbit geometry.

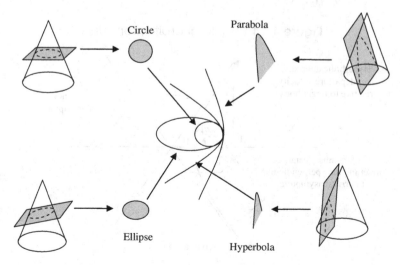

Figure 1.23 Central force system solution is a conic

Table 1.8 Properties of conic orbit

Conic	Circle	Ellipse	Parabola	Hyperbola
Eccentricity, e	0	$0 < e < 1$; $e^2 = 1 - (b/a)^2$	$e = 1$	$1 < e$; $e^2 = 1 + (b/a)^2$
Perifocal distance	a	$a(1-e)$	$p/2$	$a(1-e)$
Velocity at distance r from focus	$v^2 = \dfrac{\mu}{r}$	$v^2 = \mu\left(\dfrac{1}{r} - \dfrac{1}{a}\right)$	$v^2 = 2\dfrac{\mu}{r}$	$v^2 = \mu\left(\dfrac{2}{r} - \dfrac{1}{a}\right)$
Total energy per unit mass	$E = -\mu/2a$ < 0	$E = -\mu/2a < 0$	$E = 0$	$E = \mu/2a > 0$
Parametric equation	$x^2 + y^2 = a^2$	$(x/a)^2 + (y/b)^2 = 1$	$x^2 = 4qy$	$(x/a)^2 - (y/b)^2 = 1$

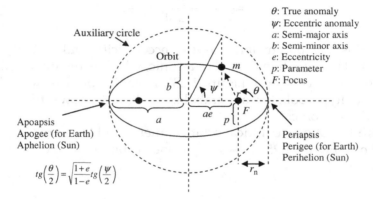

θ: True anomaly
ψ: Eccentric anomaly
a: Semi-major axis
b: Semi-minor axis
e: Eccentricity
p: Parameter
F: Focus

Auxiliary circle

Orbit

Apoapsis
Apogee (for Earth)
Aphelion (Sun)

Periapsis
Perigee (for Earth)
Perihelion (Sun)

$$tg\left(\frac{\theta}{2}\right) = \sqrt{\frac{1+e}{1-e}}\, tg\left(\frac{\psi}{2}\right)$$

Figure 1.24 Elliptical orbit geometry

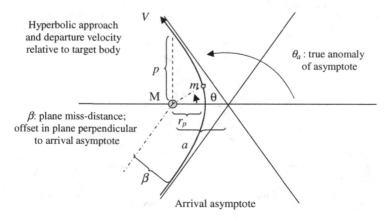

Hyperbolic approach
and departure velocity
relative to target body

θ_a : true anomaly
of asymptote

β: plane miss-distance;
offset in plane perpendicular
to arrival asymptote

Arrival asymptote

Figure 1.25 Hyperbolic orbit geometry

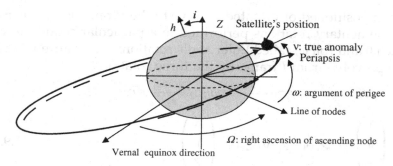

Figure 1.26   Orbital parameter

In addition to the elliptical parameters of an orbit, a few more parameters are needed in order to determine the exact location of the spacecraft, and therefore the location of the sensor. Figure 1.26 shows the reaming parameters where Ω is the longitude or right ascension of ascending node (cross point of the satellite orbit and the equatorial plane moving from south to north), measured in the reference plane from a defined reference meridian. ω is the argument of periapsis or perigee (the angle between the ascending node to the eccentricity vector measured in the direction of the satellite motion), i is the inclination of orbit plane $h \times Z$, and v is the true anomaly, the angle from the eccentricity vector to the satellite position vector.

Earth orbital patterns result from the centrifugal force of the spacecraft that balances the gravitational force. In addition to the orbital plane, spacecraft altitude gives three orbital classifications; low Earth orbit (LEO), medium Earth orbit (MEO) and geostationary orbit (GEO). Some of the advantages of the LEO orbit are the low transmit signal power, smaller delays, less station-keeping and ease of launch. However, the coverage is much smaller because of the orbit period of ~100 min, and there is more complex tracking and orbit deterioration from drag. For GEO orbits spacecraft are easier to track, with a continuous coverage. The disadvantages include a higher power signal transmission and longer delay, in addition to orbit maintenance and more limited launch vehicle options. For MEO spacecraft, they have a mix of LEO and GEO advantages and disadvantages.

1.9.4 Attitude of a Spacecraft

The general motion of a spacecraft is a combination of translation and rotation of the center of mass. One way of following how the S frame is positioned or oriented relative to the I frame is by means of three elementary rotations performed in a particular sequence: yaw ψ, pitch θ and roll ϕ, called Euler angle rotations (see Figure 1.21). Let frame S evolve from I frame by the three rotations:

$$XYZ \xrightarrow[Z,z_1]{\psi:\text{Yaw}} x_1y_1z_1 \xrightarrow[y_1,y_2]{\theta:\text{Pitch}} x_2y_2z_2 \xrightarrow[x_2,x]{\phi:\text{Roll}} xyz$$

$$\begin{pmatrix} x \\ y \\ z \end{pmatrix} = Q_x(\phi)Q_y(\theta)Q_z(\psi) \begin{pmatrix} X \\ Y \\ Z \end{pmatrix} \tag{1.9.11}$$

$$Q_x(\phi) = \begin{pmatrix} 1 & 0 & 0 \\ 0 & c\phi & s\phi \\ 0 & -s\phi & c\phi \end{pmatrix}; \quad Q_y(\theta) = \begin{pmatrix} c\theta & 0 & -s\theta \\ 0 & 1 & 0 \\ s\theta & 0 & c\theta \end{pmatrix};$$

$$Q_z(\psi) = \begin{pmatrix} c\psi & s\psi & 0 \\ -s\psi & c\psi & 0 \\ 0 & 0 & 1 \end{pmatrix}. \tag{1.9.12}$$

The angular rate of S relative to I is

$$\omega = \begin{pmatrix} 0 \\ 0 \\ \psi' \end{pmatrix}_{S1} + \begin{pmatrix} 0 \\ \theta' \\ 0 \end{pmatrix}_{S2} + \begin{pmatrix} \phi' \\ 0 \\ 0 \end{pmatrix}_S, \tag{1.9.13}$$

where the angular rate of S_2 relative to S is composed of three parts: a rate θ' about the y axis of S_2, a rate ϕ' about the x axis of S, and a rate of ψ' about the z axis of S_1. Resolving in the S frame yields:

$$(\omega)_S = Q_x(\phi)Q_y(\theta) \begin{pmatrix} 0 \\ 0 \\ \psi' \end{pmatrix}_{S1} + Q_x(\phi) \begin{pmatrix} 0 \\ \theta' \\ 0 \end{pmatrix}_{S2} + \begin{pmatrix} \phi' \\ 0 \\ 0 \end{pmatrix}_S$$

$$(\omega)_S = \begin{pmatrix} 1 & 0 & -s\theta \\ 0 & c\phi & s\phi c\theta \\ 0 & -s\phi & c\phi c\theta \end{pmatrix} \begin{pmatrix} \phi' \\ \theta' \\ \psi' \end{pmatrix} = P \begin{pmatrix} \phi' \\ \theta' \\ \psi' \end{pmatrix} \tag{1.9.14}$$

$$\begin{pmatrix} \phi' \\ \theta' \\ \psi' \end{pmatrix} = P^{-1}(\omega)_S = P^{-1} \begin{pmatrix} \omega_x \\ \omega_y \\ \omega_z \end{pmatrix}$$

$$P^{-1} = \frac{1}{c\theta} \begin{pmatrix} c\theta & s\phi s\theta & c\phi s\theta \\ 0 & c\phi c\theta & -s\phi c\theta \\ 0 & s\phi & c\phi \end{pmatrix} \neq P^T \tag{1.9.15}$$

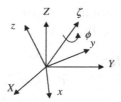

Figure 1.27   Reference frames *I* and *S* unit vectors ζ and ϕ

where P is not orthogonal. Usually ω_x, ω_y and ω_z are available using navigation sensors and ϕ, θ and ψ are to be computed. This representation is easy to visualize but has some singularity issues so that attitude control systems do not use it. Instead, quaternions are used. The following is a derivation of quaternion. Using two reference frames S and I could be related by a unit vector ζ and an angle of rotation ϕ about ζ (see Figure 1.27). ζ and ϕ are chosen such that

$$XYZ \xrightarrow[\xi]{\phi} xyz, \text{where } \xi = \begin{pmatrix} \xi_1 \\ \xi_2 \\ \xi_3 \end{pmatrix}; \xi\xi^T = 1 \tag{1.9.16}$$

$$Q = \begin{pmatrix} (1-c\phi)\xi_1^2 + c\phi & (1-c\phi)\xi_2\xi_1 + \xi_3 s\phi & (1-c\phi)\xi_3\xi_1 - \xi_2 s\phi \\ (1-c\phi)\xi_2\xi_1 - \xi_3 s\phi & (1-c\phi)\xi_2^2 + c\phi & (1-c\phi)\xi_2\xi_3 + \xi_1 s\phi \\ (1-c\phi)\xi_3\xi_1 + \xi_2 s\phi & (1-c\phi)\xi_2\xi_3 - \xi_1 s\phi & (1-c\phi)\xi_3^2 + c\phi \end{pmatrix} \tag{1.9.17}$$

$$Q\xi = \xi \Rightarrow (\xi)_S = (\xi)_I, trace(Q) = \sum_i Q_{ii} = 1 + 2c\phi;$$

$$Q = c\phi \cdot I + (1 - c\phi)\xi\xi^T - s\phi \cdot S(\xi) \tag{1.9.18}$$

For all rotation matrices, four elements, three from ζ and ϕ, could be grouped in a four by one array called quaternion. This quaternion is not a vector and does not transform as a vector.

$$\bar{q} = \begin{pmatrix} q_1 \\ q_2 \\ q_3 \\ q_4 \end{pmatrix} = \begin{pmatrix} \xi_1 s(\phi/2) \\ \xi_2 s(\phi/2) \\ \xi_3 s(\phi/2) \\ c(\phi/2) \end{pmatrix} = \begin{pmatrix} \xi s(\phi/2) \\ c(\phi/2) \end{pmatrix} = \begin{pmatrix} q \\ q_4 \end{pmatrix}. \tag{1.9.19}$$

The following are quaternion properties:

$$q_1^2 + q_2^2 + q_3^2 + q_4^2 = 1. \tag{1.9.20}$$

In order to generate the rotation matrix Q from quaternion, the following relationship could be used:

$$Q = (q_4^2 - q^T q)I + 2qq^T - 2q_4 S(q)$$

$$S(q) = \begin{pmatrix} 0 & -q_3 & q_2 \\ q_3 & 0 & -q_1 \\ -q_2 & q_1 & 0 \end{pmatrix}$$

$$q_1 = \frac{Q_{2,3} - Q_{3,2}}{4q_4}$$

$$q_2 = \frac{Q_{3,1} - Q_{1,3}}{4q_4} \tag{1.9.21}$$

$$q_3 = \frac{Q_{1,2} - Q_{2,1}}{4q_4}$$

$$q_4 = \pm\frac{1}{2}(1 + trace(Q))^{1/2}.$$

Quaternion multiplication:

$$\bar{r} = \overline{pq} = \begin{pmatrix} r_1 \\ r_2 \\ r_3 \\ r_4 \end{pmatrix} = \begin{pmatrix} p_4 & p_3 & -p_2 & p_1 \\ -p_3 & p_4 & p_1 & p_2 \\ p_2 & -p_1 & p_4 & p_3 \\ -p_1 & -p_2 & -p_3 & p_4 \end{pmatrix} \begin{pmatrix} q_1 \\ q_2 \\ q_3 \\ q_4 \end{pmatrix} \tag{1.9.22}$$

$$\left. \begin{array}{l} \bar{q}_1 = (\xi 1 \quad \phi 1) \\ \bar{q}_2 = (\xi 2 \quad \phi 2) \end{array} \right\} \Rightarrow \bar{q}_1 \cdot \bar{q}_2 = (\phi 1 \xi 2 + \phi 2 \xi 1 + \xi 1 \times \xi 2 \quad \phi 1 \phi 2 - \xi 1 \cdot \xi 2).$$

$$\tag{1.9.23}$$

Time derivative of a quaternion:

$$\frac{d\bar{q}}{dt} = \frac{\bar{\omega} \cdot \bar{q}}{2}$$

$$\bar{\omega} = (\omega 0) = \begin{pmatrix} \omega_x \\ \omega_y \\ \omega_z \\ 0 \end{pmatrix} ; \tag{1.9.24}$$

$$\frac{dq}{dt} = -\frac{S(\omega)}{2}q + \frac{q_4\omega}{2}$$

$$\frac{dq_4}{dt} = -\frac{\omega \cdot q}{2}.$$

Quaternions are used in flight computers or in simulation studies where large changes in angle are involved to track the attitude of the spacecraft. There are many advantages to using quaternions. For instance, with quaternions only four parameters are needed and Euler singularity is eliminated. Quaternions are ideal for digital control. Errors are easily checked and corrections are easily done by ensuring that the norm of the quaternion is one. However, the only disadvantage is that it is not intuitively easy to visualize a transformation or interpret it.

Bibliography

1. Alexander, M. B., Leach, R. D., and Bedingfield, K. L. (1996) *Spacecraft System Failures and Anomalies Attributed to the Natural Space Environment*, NASA Publication.
2. Blanchard, S. B. and Fabrycky, W. J. (1990) *Systems Engineering and Analysis*, Prentice Hall.
3. Boehm, B. W. (1981) *Software Engineering Economics*, Prentice-Hall.
4. Department of Defence (2003) *Parametric Estimating Handbook*, ISPA.
5. ESA (1999) *Space Engineering, Electrical and Electronic*, ECSS-E-20A.
6. ESA (1999) *Space Product Assurance*, ECSS-Q-70-04A.
7. ESA (2000) *Space Engineering, Mechanincal – Part1: Thermal Control*, ECSS-E-30 Part 1A.
8. ESA (2000) *Space Product Assurance*, ECSS-Q-70-02A.
9. ESA (2003) *Space Engineering, SpaceWire Links, Nodes, Routers and Networks*, ECSS-E-50-12A.
10. FAA (2002) *Commercial Space and Launch Insurance: Current Market and Future Outlook*, Fourth Quarter, Quarterly report.
11. Fortescue, P. and Stark, J. (1997) *Spacecraft Systems Engineering*, John Wiley and Sons.
12. Frenzel, L. (2002) *Principles of Electronic Communication Systems, Digital Principles*, McGraw-Hill.
13. Grady, J. O. (1993) *System Requirements Analysis*, McGraw-Hill.
14. Grady, J. O. (1994) *System Integration*, CRC.
15. Grady, J. O. (1997) *System Validation and Verification*, CRC.
16. ICD-GPS-200C (1993) *Navstar GPS Space Segment / Navigation User Interfaces*.
17. Johnson, F. S. (1961) *Satellite Environment Handbook*, Stanford University Press.
18. Larson, W. J. and Wertz J. R. (1992) *Space Mission Analysis and Design*, Microcosm Inc. and Kluwer Academic Publisher.
19. Martin, J. N. (1997) *System Engineering Guidebook*, CRC.
20. Mil-Std 461-C (1967) *Measurement of Electromagnetic Interference Characteristics*.
21. MIL–STD–499B (1994) *Systems Engineering*.
22. Mil-STD-1540C (1994) *Test Requirement for Launch, Upper Stage and Space Vehicle*.
23. MSFC-HDBK-670 (1991) *General Environmental Test Guidelines for Protoflight Instruments and Experiments*.

24. NASA (1998) *Electrical Grounding Architecture for Unmanned Spacecraft*, HDBK4001.
25. NASA (1996) *Payload Test Requirements*, NASA-STD 7002.
26. NASA (2001) *Dynamic Environmental Criteria*, NASA-HDBK 7005.
27. Pease, C. B. (1991) *Satellite Imaging Instruments*, Ellis Horwood Limited.
28. Pisacane, V. L. and Moore, R. C. (1994) *Fundamentals of Space Systems*, Oxford University Press.
29. Rimrott, F. P. J. (1989) *Introductory Attitude Dynamics*, Springer-Verlag.
30. Sidi, M. J. (1997) *Spacecraft Dynamics and Control, a Practical Engineering Approach*, Cambridge University Press.
31. Tribble, A. C. (1995) *The Space Environment: Implications for Spacecraft Design*, Princeton University Press.
32. Wertz, J. R. (1978) *Spacecraft Attitude Determination and Control*, Reidel Publishing Company.
33. Wertz, J. R. and Larson, J. W. (1996) *Reducing Space Mission Cost*, Microcosm Press and Kluwer Academic Publishers.

2

Sensors and Signals

If something exists, it can be detected and measured. However, challenges reside in the approach to conducting the measurement. A signal can be emitted, received and manipulated as long as the process is in place for that signal or information to reach a sensor. A sensor amplifies and translates the signal into a number or value that has a meaning for interpretation. In this chapter, sensor characteristics and various types of signals are explored. Various techniques for enhancing signals, interpreting information and processing information will be presented.

2.1 Sensor Characteristics

There are many ways to characterize a sensor. There are both static and dynamic aspects that can be examined. The static characteristics of a sensor include the steady state regime and can determine accuracy and resolution. These include spatial resolution, bandwidth, range, sensitivity, linearity and hysteresis. Dynamic characteristics of a sensor are determined by analyzing the response to an impulse, step or ramp function. Each sensor has some limitation in measurement. There are many factors that could induce a systematic error such as interferences between variables, drift or complications during the transmission process. Compensation methods such as feedback and filtering could correct for some systematic errors. The

Spacecraft Sensors. Mohamed M. Abid
© 2005 John Wiley & Sons, Ltd

knowledge of errors allows us to obtain the most probable value from a set of scattered observations, assess the accuracy of a set of measurements and determine if they meet the required or specified accuracy.

2.1.1 Accuracy and Precision

Accuracy is the capability of a measuring instrument to give results close to the true value of the measured quantity. Accuracy could also be defined as the maximum deviation of measurement from the true value. It is the difference between the true value of a measured quantity and the most probable value which has been obtained from measurement. Inaccuracy is measured by the absolute and relative errors. Absolute error is the difference between the measurement and the true value. Relative error is the ratio of the absolute error to the true value. It is measured as the highest deviation of a value returned by the sensor from the true value at its input.

Using calibration, accuracy could be enhanced up to, but not exceeding, the instrument precision. Resolution is the smallest increment of output that a measuring system can sense which is a characteristic of the measuring instrument. The uncertainty due to the resolution of any instrument is half of the smallest increment displayed. For example, with a 1 mm ruler increment, a typical measurement would be 5 mm +/−0.5 mm.

Precision is a gauge of how close the measured values are when multiple measurements are conducted. It is a measure of repeatability, i.e. the degree of agreement between individual measurements of a set of measurements, all of the same quantity. Therefore precision is a necessary, but not sufficient, condition for accuracy. It is affected by the measuring technique. An example of a measurement which is precise but not accurate is a tape-measure which is stretched. This induces a systematic error. Figure 2.1 shows a graphical representation of accuracy and precision.

Table 2.1 shows expressions for the accuracy, uncertainty, bias and error of a value where x is the true value and x_i is the measured value. The domain of a sensor is determined by the minimum and the maximum acceptable value. The range is determined by the minimum and the maximum reachable value of the sensor's output. The dynamic range D_r of a sensor, which is one of the essential features defining the precision of the measurement, is the ratio between the

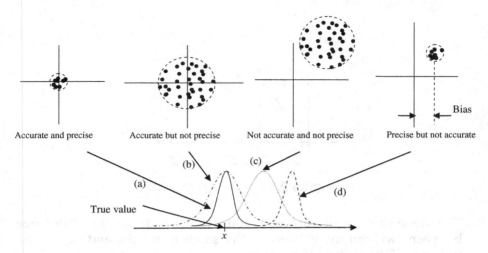

Figure 2.1 Signal (a) is accurate and precise, (b) is accurate but not precise, (c) neither accurate nor precise, (d) precise but not accurate

largest admissible values s_m by the sensor input, and the largest increase in the input s_r without modifying the output

$$D_r \equiv s_m/s_r. \qquad (2.1.1)$$

Precision increases when D_r increases.

2.1.2 Hysteresis

A sensor is said to exhibit hysteresis when there is a difference in readings depending on how the measurement was approached, i.e. from a higher input or from a lower input (see Figure 2.2). Hysteresis can result from mechanical, thermal, magnetic or other effects and can be expressed:

$$\text{hysteresis} = |y_1 - y_2|_{max}/|x_{max} - x_{min}|. \qquad (2.1.2)$$

Table 2.1 Accuracy, uncertainty, bias and error expressions

Accuracy	Uncertainty	Bias	Error
$|x - x_i|_{max}$	$|\bar{x} - x_i|_{max}$	$x - \bar{x}$	$x - x_i$

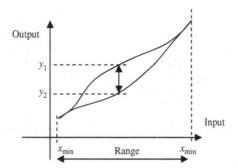

Figure 2.2 Typical hysteresis

The denominator of this equation corresponds to the difference between two output values for the same input depending on the trajectory followed by the sensor.

2.1.3 Calibration

A sensor's behavior could differ from one environment to another. Calibrating a sensor is a means of normalizing the sensor response to the environment that is not of interest for the measurements. Calibration could be conducted at many different points during manufacturing, integration, operation and between measurements. It is a reality check and ensures an expected behavior of the sensor. The classic approach to calibration would be to measure a known physical input and then measure the response and compare it to a known output. The drift in the measurement, resulting from cumulative error, or not compensating for small factors, could easily be identified and quantified when calibration is conducted. Calibration establishes the accuracy of the instruments. Rather than taking the reading of an instrument as is, it is usually best to make a calibration measurement to validate it. Careful calibration should be conducted because improper calibration of the sensors can lead to bias and systematic errors, where effects cannot be reduced by observation and repetition.

Pre-flight calibrations are conducted on most sensors in order to measure their main performance characteristics. However, the degradation of the spacecraft due to the environment, noise level variation, power and other factors during its life duration on orbit requires a routine in-flight recalibration of the sensor. For example the dark current, which is the basic flow of current without any external stimulus, in an ensemble of IR sensors can vary during the mission because of contamination. Thus, in order to normalize each IR sensor reading, one should perform a periodic reading of a known configuration such as the blocking of all light from entering the detectors.

2.1.4 Transfer Function

The static characteristics of a sensor are described by the sensor's response to a constant input. When the input is variable, the sensor's response becomes dynamic. This could mean that the system has an energy storage element. Dynamic characteristics are determined by analyzing the response of the sensor to a set of input variables such as an impulse, step or sinusoid variable. A system can be described in terms of $x(t)$ written in a differential equation

$$a_n \frac{d^n y}{dt^n} + a_{n-1} \frac{d^{n-1} y}{dt^{n-1}} + \cdots + a_1 \frac{dy}{dt} + a_0 y = x(t) \qquad (2.1.3)$$

where $x(t)$ is the input or the forcing function imposed on the system. The order of the system is defined by the order (n) of the differential equation. A zeroth-order system takes the following form:

$$a_0 y = x(t). \qquad (2.1.4)$$

The term $1/a_0$ is called the static sensitivity. Zero-order systems do not include energy storing elements. This system is the desired response of a sensor since there are no delays, and the bandwidth is infinite. However, it is far from realizable because of instrument limitations. This is more like the ideal sensor. First-order system sensors have one element that stores energy and one that dissipates it. A first-order system takes the following form:

$$a_1 \frac{dy}{dt} + a_0 y = x(t). \qquad (2.1.5)$$

A second-order system, such as a spring mass damper accelerometer, would be:

$$a_2 \frac{d^2 y}{dt^2} + a_1 \frac{dy}{dt} + a_0 y = x(t). \qquad (2.1.6)$$

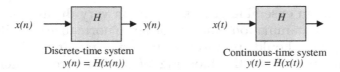

Figure 2.3 Transfer function for discrete and continuous-time systems

These dynamic models are often analyzed using the Laplace transform which has the advantage of converting differential equations into polynomial expressions. For a system that is formed by an agglomeration of subsystems, it is very often identified by a transfer function. This is a very convenient way of analyzing a system without knowing much about the intricacies of the subsystems. Each subsystem will have a specific transfer function. Once a signal or signals are input into the system, the output can be analyzed independently for each subsystem. Figure 2.3 shows a typical representation of a transfer function block.

A sensor could be seen as a box that transforms input signals to output signals. Combinations of discrete and analog signals are possible using analog-to-digital converters and digital-to-analog converters. A sensor could also be represented as an ensemble of sub boxes set in a variety of configurations. The more complex the system is, the less one wants to know the intricacy of the design (see Figure 2.4 for an illustration of the transfer function of different configurations of subsystems).

2.2 Types of Signals

A signal is a physical quantity that transports a message (m). It could be sent or received. The general representation of a signal

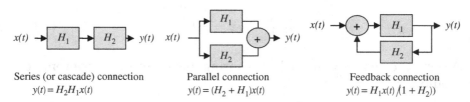

Figure 2.4 Various combinations of systems with the total transfer function

is the following

$$s \equiv s(t, m, f_c, \phi, A, \dots) \tag{2.2.1}$$

where the variables could be t for time, m for message, ϕ for phase, f_c for carrier frequency, and A for amplitude. The name of the signal depends on its nature, and there are many types. A deterministic signal is completely specified as a function of time. A signal may be mathematically expressed as a function of time, $x(t)$, in which the signal is periodic.

The schematic in Figure 2.5 shows a typical classification of signals. A deterministic dynamic environment is reproducible within the experimental error. It is also predictable based on prior measurement. A random environment is said to be non-stationary if the signals representing the environment are ongoing, but have at least one average property that varies with time. A signal is said to be transient if the signal has a clear beginning and end, and a relatively short duration compared with the decay time of the impulse response function. Continuous and discontinuous signals are similar to fractals, the more you inspect the signal the more you discover the random aspect of the signal or the missing point. The nature of a signal could vary from one domain to another. A continuous signal can be seen as discrete if it is scrutinized. Ergodic signals are signals where all time averages of any sample function are equal to the corresponding ensemble averages. They are always stationary.

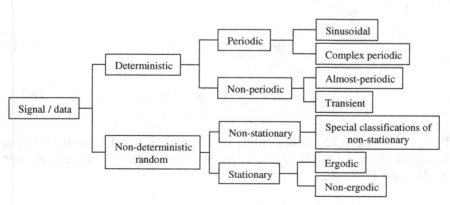

Figure 2.5 Typical signal classification[2]

2.2.1 Signal Properties

Sensor specifications are derived based on the property of the measured signal. The knowledge of some basic signal properties such as its power would help in designing the sensor. There are many properties for a signal that one can use. Mathematically, the difference between continuous and discontinuous is in the summation process. The time-average value of a signal $\langle s \rangle$, the root mean square (RMS) value of a signal s_{RMS}, the energy E and the power P of a signal for continuous signal are defined:

$$\bar{s} = \langle s \rangle = \lim_{T \to \infty} \frac{1}{T} \int_{-T/2}^{T/2} s(t)dt \tag{2.2.2}$$

$$s_{RMS} = \left[\lim_{T \to \infty} \frac{1}{T} \int_{-T/2}^{T/2} s^2(t)dt \right]^{1/2} \tag{2.2.3}$$

$$E = \lim_{T \to \infty} \int_{-T/2}^{T/2} |s(t)|^2 dt = \lim_{T \to \infty} \int_{-T/2}^{T/2} s(t) \cdot s(t)^* dt, \tag{2.2.4}$$

where the * stands for the conjugate, and

$$P = \lim_{T \to \infty} \frac{1}{T} \int_{-T/2}^{T/2} |s(t)|^2 dt. \tag{2.2.5}$$

For a discrete signal, the energy and power are

$$E = \lim_{N \to \infty} \sum_{n=-N}^{N} |s(n)|^2 \tag{2.2.6}$$

$$P = \lim_{N \to \infty} \frac{1}{2N+1} \sum_{n=-N}^{N} |s(n)|^2. \tag{2.2.7}$$

Because of the wide range that power can take and because adding is easier than multiplying, it is often convenient to express power in dB (see Table 2.2)

$$S_{dB} = 10 \log_{10}(S) \Rightarrow S = 10^{S/10}. \tag{2.2.8}$$

Table 2.2 Conversion between watts, dBm and dBW

Watts	0.001	0.002	0.01	0.1	1	2	10	100
dBm	0	3.0103	10	20	30	33.01	40	50
dBW	−30	−26.99	−20	−10	0	3.0103	10	20

Also, dBc is sometimes used. dBc is the power of one signal refer-enced to a carrier signal, i.e. if a second harmonic signal at 10 GHz is 3 dB lower than a fundamental signal at 5 GHz, then the signal at 10 GHz is −3 dBc.

2.2.2 Periodic Signals

Periodic signals are signals that satisfy the periodicity criterion, i.e. for continuous signals with period T:

$$s(t) = s(t + T) \Rightarrow s(t) = s(t + kT); \quad k = 0, 1, 2, \ldots . \qquad (2.2.9)$$

For a discrete periodic signal with period N:

$$s(n) = s(n + N) \Rightarrow s(n) = s(n + kN); \quad k = 0, 1, 2, \ldots . \qquad (2.2.10)$$

Most electromagnetic sensors exploit the properties of sinusoidal signals. In the time domain, such signals are constructed of sinu-soidally varying voltages or currents constrained within wires. The time-average value of a periodic signal is

$$\bar{s} = \lim_{T \to \infty} \frac{1}{T} \int_0^T A \cos(2\pi f t + \phi) dt$$

$$\bar{s} = \frac{A}{T} \left(\cos(\phi) \int_0^T \cos 2\pi f t \, dt - \sin(\phi) \int_0^T \sin 2\pi f t \, dt \right) = 0. \qquad (2.2.11)$$

The power-average value of a periodic signal is

$$P = \lim_{T \to \infty} \frac{1}{T} \int_0^T A^2 \cos^2(2\pi f t + \phi) dt = \frac{A^2}{2}. \qquad (2.2.12)$$

2.2.3 Representing Signals as Impulses

It is possible to regenerate an arbitrary signal by sampling it with a shifted unit impulse. For a system H, the response of that system to

Figure 2.6   Response of a system to delta function input

an input signal of unit impulse $\delta(t)$ is called its impulse response, and is represented by $h(t)$. The impulse function has the following properties:

$$\delta(-t) = \delta(t)$$
$$x(t)\delta(t - t_0) = x(t_0)\delta(t - t_0)$$
$$\delta(at) = \delta(t)/|a|$$
$$\int_{-\infty}^{+\infty} x(t)\delta(t - t_0)dt = x(t_0)$$
$$\int_{-\infty}^{+\infty} x(t)\delta^n(t - t_0)dt = (-1)^{n-1}x(t_0). \qquad (2.2.13)$$

For the continuous case, $h(t) = H(\delta(t))$ (see Figure 2.6) For the discrete case substitute t with n.

Therefore, by knowing the sensor's response to an impulse function one can virtually reconstruct the response to most signals by using convolution properties as shown in Figure 2.7.

An input function x can be written in terms of δ as follows:

$$x(n) = \sum_{k=-\infty}^{\infty} x(k)\delta(n - k). \qquad (2.2.14)$$

The output signal y is

$$y(n) = H(x(n)) = H\left(\sum_{k=-\infty}^{\infty} x(k)\delta(n - k)\right) \qquad (2.2.15)$$

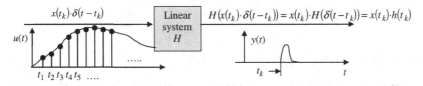

Figure 2.7   Convolution

where x is the input, H is the transfer function and h is the impulse response function. For linear time invariant (LTI) systems which are additive, homogeneous and invariant,

$$y(n) = \sum_{k=-\infty}^{\infty} x(k)h(n-k).\qquad(2.2.16)$$

$y(n)$ is called the convolution sum or superposition sum and is represented by

$$y(n) \equiv x(n) \otimes h(n) \neq x(n) \cdot h(n).\qquad(2.2.17)$$

The product sign for the convolution is different from the dot product sign. In the continuous domain, the convolution is defined as

$$f(t) \otimes g(t) = \int_{-\infty}^{+\infty} f(x) \cdot g(t-x)dx.\qquad(2.2.18)$$

The response to an impulse function is

$$f(t) \otimes \delta(t-a) = \int_{-\infty}^{+\infty} f(t-x) \cdot \delta(x-a)dx = f(t-a).\qquad(2.2.19)$$

This equation shows that the convolution with a delta function shifts the function. Overall, a sensor would be subject to a signal that could be broken into a set of impulse signals. Impulses could be used to determine the sensor characteristics. They can also be used as a calibration source for the sensor in order to capture most of the signal that a sensor is anticipated to measure.

2.2.4 Random Signals

One of the main challenges in processing a signal is being able to distinguish between signal and noise or extract the signal which is embedded in a random signal (Figure 2.8). The presence of randomness in the sensing element could be the result of the measured phenomenon, or just an interaction with a noisy source. Not knowing the governing equation, a non-deterministic signal could be probabilistically characterized and analyzed.

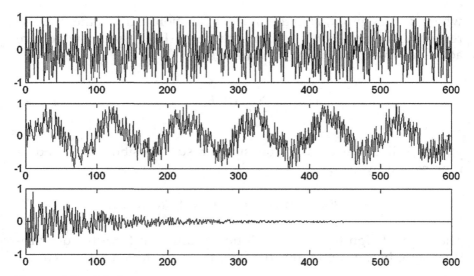

Figure 2.8 (a) Random signal, (b) random plus periodic function, (c) random transient signal[†]

Intrinsically random data or phenomena cannot be described by a deterministic mathematical relationship. Any measurement or observation represents only one of many possible results which might have occurred. In other words, only a statistical description is possible. A sampling function is a single time representation of a random phenomenon. A random process or a stochastic process is the collection of all possible sample functions.[9] For a fragment signal $s(t, r)$ where t is time and r is a random variable, if r is fixed, $s(t)$ is called a sampling function. If t is fixed, $s(t, r) = s(t)$ is called a random variable. The random variable $s(t)$ can be discrete or continuous.

Characterization of a Random Variable

There are mainly four statistical functions that describe the basic properties of a random signal. The mean square value, Ψ_s^2, which is the average of the squared values of the time history, gives a rudimentary description of the intensity of the signal. The RMS value, ψ_s, which describes a combination of the central tendency and dispersion, is the square root of the mean square value. The mean value, μ_s,

---

[†]Figure 2.8 was generated using the following Matabla code. For $t=1:600$; $y0(t)$=rand* (-1) ^round (rand); $y1(t)$=rand*(-1)^round(rand)+sin(2*pi*t/100); y2(t)=exp$(-.01$*t)*(-1)^round(rand)*rand; end subplot(1,3,1);plot(y0); subplot(1,3,2);plot(y1/max(y1)); subplot(1,3,3);plot(y2)

describes the central tendency of $s(t)$. The variance is the square value about the mean. The standard deviation σ_s describes the dispersion of $s(t)$. For steady state signals represented by the signal $s(t)$,

$$\mu_s = \frac{1}{T}\int_0^T s(t)dt$$

$$\sigma_s = \sqrt{\frac{1}{T}\int_0^T [s(t) - \mu_s]^2 dt}$$

$$\psi_s = \sqrt{\frac{1}{T}\int_0^T s^2(t)dt}$$

$$\sigma_s^2 = \psi_s^2 - \mu_s^2.$$

(2.2.20)

Probability Density Function

The probability density function (PDF) gives information regarding the properties of the signal in the amplitude domain. This identifies the power distribution. The PDF of a random variable describes the probability that the data will assume a certain value within some defined range at any instant of time. The principal applications of the PDF are to establish a probabilistic description of a signal, or to distinguish between two signals. To find the probability of a random variable taking on a value between x_1 and x_2, the density function is integrated between these two values.

For the sample time history represented in Figure 2.9, the probability that $s(t)$ is between s and $s + \Delta s$ is

$$P(s < s(t) < s + \Delta s) = \sum_i \Delta t_i / T$$

(2.2.21)

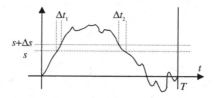

Figure 2.9 Sample time history

where T is the duration of the signal. For small Δs the PDF p is

$$p(s) = \lim_{\Delta s \to 0} \frac{P(s < s(t)s + \Delta s)}{\Delta s} = \lim_{\Delta s \to 0} \sum_i \Delta t_i / \Delta s \cdot T \geq 0. \qquad (2.2.22)$$

When the probability distribution is known, the following statistical descriptions are used for a random variable s with density function $p(s)$:

$$\mu_s = \int s \cdot p(s) ds$$
$$\sigma^2 = \int [s - \mu_s]^2 p(s) ds \qquad (2.2.23)$$

where μ_s is the mean value or the expectation, σ^2 is the variance and σ is the standard deviation. The variance represents the spread about the mean. Knowing the standard deviation and the mean of a random variable is not enough to fully characterize the random variable. The probability density function, however, is usually sufficient to characterize a random variable. This is true as well for cumulative distribution functions. Note that a numerical value of $p(s)$ is not a probability of s.

The probability that s takes a value between a and b (Figure 2.10) is

$$P(a \leq X \leq b) = \int_a^b f(x) dx. \qquad (2.2.24)$$

The total area under the graph is equal to one. Some PDFs of discrete random variables are binomial, Poisson and geometric function. Examples of continuous random variables are the uniform distribution and the Gaussian distribution.

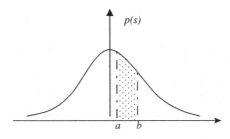

Figure 2.10 The probability that s takes a value between a and b

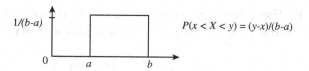

Figure 2.11 Example of a PDF

A continuous uniform distribution is defined on an interval from a to b such that the probability of x is $1/(b-a)$ for x between a and b and zero elsewhere (see Figure 2.11). Since this is the graph of a PDF, the area of the rectangle is one. An equation for the case of an exponential distribution commonly used in reliability analysis is

$$f(t) = \lambda e^{-\lambda t}, t \geq 0. \qquad (2.2.25)$$

The PDF is then

$$P(t \leq T) = 1 - e^{-\lambda t}. \qquad (2.2.26)$$

This equation represents the Weibull distribution which is used in reliability modeling and lifetime modeling. The distribution of the lifetime of components which consist of many parts, and fail when the first of these parts fail is

$$f(x) = \frac{\beta}{\mu} \left(\frac{x}{\mu}\right)^{\beta-1} e^{-\left(\frac{x}{\mu}\right)\beta}. \qquad (2.2.27)$$

This Weibull distribution is important because it gives a good representation of the frequency distribution of failures of many types of equipment. As for the Gamma distribution, μ and β define the characteristic shape of the Weibull PDF curve.

The Gaussian distribution is also known as the normal distribution. It is a very common distribution that describes many random variables or signals. This distribution has many attractive properties such as any linear operation on a Gaussian distribution is a Gaussian distribution. Also, the distribution of the sum of a large number of identically distributed random variables could be approximated by a Gaussian distribution. This property is known as the central limit theorem. The normal or Gaussian probability function is very important. In the one-dimensional case, it is described as a bell-shaped curve, and it is defined by a probability density function

$$p(x) = \frac{1}{\sqrt{2\pi}\sigma} \exp\left(-\frac{1}{2}\left(\frac{x-\mu}{\sigma}\right)^2\right) \qquad (2.2.28)$$

where μ is the mean and σ is the standard deviation. The width of the curve at the points of inflexion σ is used for measuring precision. A smaller width yields more precise measurements. If a series of measurements of independent values are added, the standard deviation of the result is given by

$$\sigma_{\text{sum}} = \sqrt{\sum_{i=1}^{n} \sigma_i} \qquad (2.2.29)$$

where n is the number of values added. If each has the same standard deviation σ,

$$\sigma_{\text{sum}} = \sqrt{n}\sigma.$$

The shaded area in Figure 2.12 represents 68.3% of the total area. Therefore the standard deviation σ of a series of measurements indicates a 68.3% probability that the true value for the measurement lies within $+\sigma$ or $-\sigma$ of the mean value of the data set ($2\sigma = 95.4\%$ and $3\sigma = 99.7\%$).

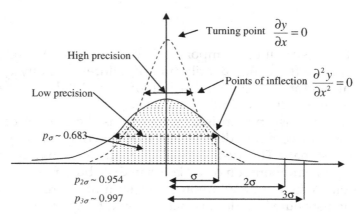

Figure 2.12 Gaussian distribution with typical approximation used in the PDF evaluation

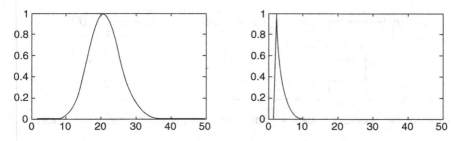

Figure 2.13 Poisson and Geometrion distribution[†]

The Poisson distribution (see Figure 2.13), is appropriate for applications that involve counting the number of times a random event occurs in a given amount of time, distance or area. λ is both the mean and the variance of the distribution

$$s(k) = \lambda^k e^{-\lambda}/k!; \quad k \in \aleph. \tag{2.2.30}$$

Equation (2.2.31) represents the geometric distribution which is mainly used for modeling the runs of consecutive successes, or failures, in repeated independent trials of a system. The geometric distribution models the number of successes before the occurrence of one failure in an independent succession of tests where each test results in success or failure

$$s(k) = (1 - p)^{k-1} p; k \in \aleph. \tag{2.2.31}$$

Correlation and Auto-correlation Function

Correlation is the process of comparing two signals, whether for radar systems, telemetry transfer communication or for calibration. Given two signals s_1 and s_2, the correlation function or the cross-correlation R_{12} is

$$R_{12}(\tau) = \lim_{T \to \infty} \frac{1}{T} \int_0^T s_1(t) s_2(t - \tau) dt \tag{2.2.32}$$

[†]Figure 2.13 was generated using the following Matlab code. n=60;l=20;p=0.5; for k=1:n; sp(k+1)= l^k*exp(l)/factorial(k); sg(k+1)=(1−p)^(k−1)*p;end subplot(1,2,1);plot (sp/max(sp)); subplot(1,2,2); plot(sg/max(sg))

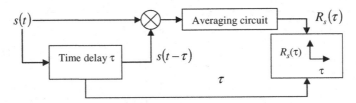

Figure 2.14 Function block diagram of an auto-correlation analyzer[2]

where τ is the searching parameter. The larger R_{12} is, the more the two signals agree. The auto-correlation function provides a tool for detecting deterministic data which might be masked in a random background. It is the correlation of the same signal. The auto-correlation between the values of $s(t)$ and $s(t+\tau)$ is obtained by taking the product of the two values and averaging over the observation time T which leads to the auto-correlation function as T approaches infinity:

$$R_s(\tau) = \lim_{T \to \infty} \frac{1}{T} \int_0^T s(t)s(t-\tau)dt. \qquad (2.2.33)$$

In general,

$$\mu_s = \sqrt{R_s(\infty)}$$
$$\psi_s = \sqrt{R_s(0)} \qquad (2.2.34)$$

where μ_s is the mean. Figure 2.14 represents a typical function block diagram of an auto-correlation analyzer.

This diagram illustrates the sequence where the signal is originally delayed by τ seconds, then multiplies the signal value at any instant by the value that had occurred τ seconds before and finally averaging the instantaneous product value over the sampling time.

Power Spectral Density (PSD)

The principal application of the PSD is to establish the frequency composition of the data which characterizes the physical system involved. Once identified, a filtering technique could be used by eliminating a specific frequency. The PSD describes the general frequency composition of the signal in terms of the spectral density of its mean square value

$$\psi_s^2(f, \Delta f) = \lim_{T \to \infty} \frac{1}{T} \int_0^T s^2(t, f, \Delta f) dt, \qquad (2.2.35)$$

the mean square value of a sample time history record in frequency range between f and $f + \Delta f$. The power spectral density function $G_s(f)$:

$$G_s(f) = \lim_{\Delta f \to 0} \frac{1}{\Delta f} \left(\lim_{T \to \infty} \frac{1}{T} \int_0^T s^2(t, f, \Delta f) dt \right). \qquad (2.2.36)$$

The power spectral density function is related to the auto-correlation function. For stationary signals, G_s and R_s are related by a Fourier transform as follows

$$G_s(f) = 2 \int_0^\infty R_s(\tau) e^{-j2\pi f \tau} d\tau t = 4 \int_0^\infty R_s(\tau) \cos(2\pi f \tau) d\tau. \qquad (2.2.37)$$

The test for periodicity in the data will theoretically appear as delta functions in the power spectrum of the data. Figure 2.15 represents a typical block diagram of a PSD analyzer.

The signal in Figure 2.15 is filtered by a band-pass filter of bandwidth Δf, and then the signal is squared and averaged over the sampling time. The mean square output is then divided by the bandwidth. The PSD is plotted as a function of the frequency.

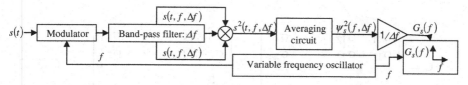

Figure 2.15 Typical PSD block diagram[2]

Covariance

For large systems of random variables, such as GPS range measurements and position estimates, the variances and covariance are arranged in a matrix called the variance–covariance matrix or the covariance matrix:

$$\mu = E(x) = \begin{pmatrix} E(x_1) \\ E(x_2) \\ \vdots \\ E(x_d) \end{pmatrix} = \begin{pmatrix} \mu_1 \\ \mu_2 \\ \vdots \\ \mu_d \end{pmatrix} = \sum_x xP(x) \tag{2.2.38}$$

where E is the expected value. The covariance matrix C is defined as the square matrix

$$C = E\left((x - \mu)(x - \mu)^T\right) \tag{2.2.39}$$

whose ijth element σ_{ij} is the covariance of x_i and x_j:

$$\mathrm{cov}(x_i, x_j) = \sigma_{ij} = E((x_i - \mu_i)(x_j - \mu_j)); \quad i,j = 1,2...d. \tag{2.2.40}$$

Using the distributive and commutative properties of the convolution function one can derive the Cauchy–Schwartz inequality

$$\sigma_{xy}^2 \le \sigma_x^2 \sigma_y^2. \tag{2.2.41}$$

The correlation coefficient is the normalized covariance:

$$\rho(x,y) = \sigma_{xy}/(\sigma_x \sigma_y)$$
$$-1 \le (x,y) \le 1. \tag{2.2.42}$$

The variables x and y are uncorrelated if $\rho = 0$.

$$C = \begin{bmatrix} \sigma_{11} & \sigma_{12} & \cdots & \sigma_{1n} \\ \sigma_{12} & \sigma_{22} & \cdots & \sigma_{2n} \\ \vdots & \vdots & \ddots & \vdots \\ \sigma_{1n} & \sigma_{2n} & \cdots & \sigma_{nn} \end{bmatrix} = \begin{bmatrix} \sigma_1^2 & \sigma_{12} & \cdots & \sigma_{1n} \\ \sigma_{12} & \sigma_2^2 & \cdots & \sigma_{2n} \\ \vdots & \vdots & \ddots & \vdots \\ \sigma_{1n} & \sigma_{2n} & \cdots & \sigma_n^2 \end{bmatrix}. \tag{2.2.43}$$

A covariance matrix is symmetric. Its diagonal elements are positive and usually non-zero since these are variances. Its eigenvalues are all positive or null. Given the covariance matrix of a set of random variables, the characteristics of expected values can be used to determine the covariance matrix of any linear combination of the measurements. If the variables are statistically independent, the covariances are zero, and the covariance matrix is diagonal. Propagation of variances is used extensively to estimate the statistical character of quantities derived by linear operations on random variables. Because of the central limit theorem, if the random variables used in the linear relationships have a Gaussian distribution, then the resultant random variables will also have a Gaussian distribution. For instance

$$\left.\begin{matrix} C = \begin{pmatrix} \sigma_1^2 & \sigma_{12} \\ \sigma_{12} & \sigma_2^2 \end{pmatrix} \\ y = x_1 \pm x_2 \end{matrix}\right\} \Rightarrow \sigma_y^2 = \sigma_1^2 + \sigma_2^2 \pm 2\sigma_{12} \qquad (2.2.44)$$

where x_1 and x_2 are random variables. Notice here that depending on the sign and magnitude of σ_{12}, the sum or difference of two random variables can either be very well determined when the variance is small or poorly determined if the variance is large. If σ_{12} is zero, uncorrelated in general and independent for Gaussian random variables, then the sum and difference have the same variance. Even if the covariance is not known, then an upper limit can be placed on the variance of the sum or difference since the value of σ_{12} is bounded.

2.3 Transforming a Signal

Typically a phenomenon would have a signature, or a signal characteristic. A sensor would be designed to identify that signature, extract it, isolate it and interpret it. In this sequence, and because of the nature of available processing, the signal is transformed. Progress in computation techniques made analog to digital transformation a standard procedure for signal processing. Figure 2.16 illustrates the sequence of transformation that an analog signal undergoes from emittance all the way to processing.

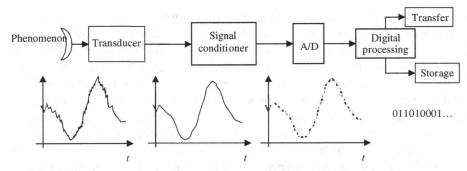

Figure 2.16   A typical sequence for sensor's signal processing[15]

2.3.1 Analog-to-digital Converter

The input signal of most sensors is in analog form. The processing, however, is inherently digital in order to benefit from the flexibility in analysis and storage. There is also better control over accuracy requirements and reproducibility with digital processing. This digitization process, however, may present some limitations such as the (analog-to-digital (A/D) converter speed and processor speed, as wide-band signals are still difficult to treat in real-time systems. There are two separate and distinct operations in the digitization process.[15] First is sampling, which is the process of defining the instantaneous points at which the signals are to be observed, and second is quantization, which is the conversion of input values at the sampling points into numerical form. The problem with sampling is determining the appropriate sampling interval τ. Over-sampling occurs for small τ, which is not a detriment unless data storage or data handling processing is used, since computation costs will increase. Aliasing occurs when τ is large. The cut-off frequency f_c quantifies this large and small τ

$$f_c = 1/2\tau. \tag{2.3.1}$$

For $f = 100$ Hz, τ less than 0.005 is sufficient. If there is a reason to believe that the data may contain contributions up to 200 Hz, then τ should be less than 0.0025 in order to avoid aliasing. A rule of thumb would be to make sure f_c is twice as large as the maximum frequency anticipated. This is the Nyquist criterion.

An analog-to-digital converter (ADC) takes an analog signal as an input and then outputs a digital signal at each sample time. Sampling

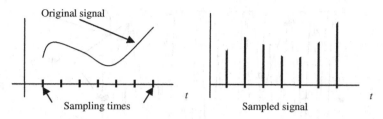

Original signal

Sampling times

Sampled signal

Figure 2.17   Sampled signal[11]

may be continuous but is usually done at a specified interval (see Figure 2.17). One trade-off in this procedure is that the converter will be limited by the conversion time, the required time before the converter can provide valid output data and by the converter throughput rate, which is the number of times the input signal can be sampled maintaining full accuracy. It is also the inverse of the total time required for one successful conversion.

Sampling of a continuous-time signal $x(t)$ can be done by obtaining its values at periodic times kT where T is the sampling period. Conceptually this can be done by multiplying $x(t)$ by a train of impulses with a period. A mathematical description of the sampling process may be obtained using an impulse modulator model:

$$x_s(t) = x(t)p(t)$$

$$p(t) = \sum_{k=-\infty}^{\infty} \delta(t - kT) \qquad (2.3.2)$$

$$x_s(t) = \sum_{k=-\infty}^{\infty} x(kT)\delta(t - kT).$$

To ensure accurate reconstruction of the signal the Nyquist sampling criterion should be satisfied. Nyquist states that the sampling rate of a signal should be at least twice the highest frequency component in order to reconstruct the signal accurately from the sampled one. For example, the sampling frequency of the signal in Figure 2.17 should be larger than 500 Hz:

$$s(t) = 3\cos(5\pi t) + \sin(30\pi t) - \cos(250\pi t) \Rightarrow f_S > 500 \text{ Hz}. \qquad (2.3.3)$$

Figure 2.18 shows that when the sampling frequency is less than the Nyquist frequency, the original analog signal $s(t)$ cannot be recovered from $s_s(t)$ since there will be unwanted overlaps in $S_s(\omega)$. This

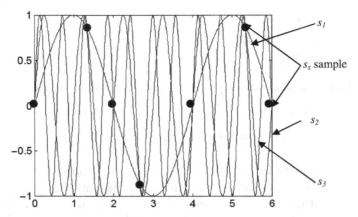

Figure 2.18 Reconstruction of sampled function with three functions with different frequencies[†]

will cause the recovered signal $s_r(t)$ to be different from $s(t)$. This is called an aliasing or under-sampling.

In the frequency domain, an analog signal may be represented in terms of its amplitude and total bandwidth. A sampled version of the same signal can be represented by a repeated sequence spaced at the sample frequency f_s. If the sample rate is not sufficiently high, then the sequences will overlap and high frequency components will appear at a lower frequency. Over-sampling is the process of sampling at frequencies $f_s >> 2f_{max}$. Over-sampling and averaging may improve ADC resolution, but will increase calculations as well as memory allocation.

Quantization is the process of taking an analog signal and converting it into a digital signal. Therefore the original signal is approximated at discrete levels. The vertical resolution depends on the number of levels based on the required accuracy. The horizontal accuracy is defined by the sampling rate which is dependent on the signal frequency and should satisfy the Nyquist criterion.

The converter is based on two ranges of the input which are typically voltage and the number of bits in the output. The quantization step Q or the resolution of the least significant bit (LSB) is defined as

$$Q = (V_{max} - V_{min})/(2^n - 1) \qquad (2.3.4)$$

[†]Figure 2.18 was generated using the following matlab code. $t=0:0.01:6; f0=0.5; s1=\sin(1*3.14*f0*t);$ plot($t,s1$);hold on; $s2=\sin(4*3.14*f0*t)$;plot($t,s2$);hold on;$s3=\sin(7*3.14*f0*t)$;plot($t,s3$);

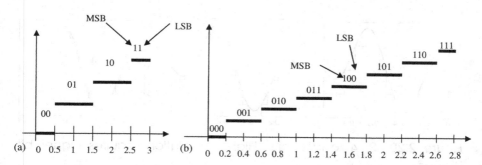

Figure 2.19 (a) A four-level of quantization transfer characteristic using 2 bits, 3V range using 2 bits, $Q = 1$ V per division. (b) An eight-level of quantization transfer characteristic using 3 bits, 2.8 V range using 3 bits, $Q = 0.4$ V per division[15] (the error is 0.2 V/div)[†]

where V_{max} is the maximum value of V and V_{min} is the minimum value of V, n is the number of output bits. For instance, for $V_{max} = 3$ V, $V_{min} = 0$ V and the output is 2 bits (see Figure 2.19). The quantization interval, which is also referred to as the resolution, is $Q = (3 \text{ V}–0 \text{ V})/(2^n – 1)$. Therefore the maximum error in an input supplies is $\pm Q/2 = \pm \frac{1}{2}$.

One can see from Figure 2.19 that a given digital output corresponds to a range of possible input's, and that the smallest change in the digital output is equal to the weight of the least significant bit (LSB). The input voltage is continuous, which means that a given signal output must correspond to a range of input values, i.e. quantization. For instance, for a voltage range of 10V, 8 bits will give a resolution of 39 mV, for 10 bits the resolution is 9.8 mV and for 16 bits the resolution is 153 μV. Figure 2.20 illustrates the importance of the number of bits in signal reconstruction.

Typical parameters of interest in the process of ADC are the numbers of bits, N, which are related to the resolution, the data throughput which determines the processing rate, the signal-to-noise ratio (SNR) and spurious-free dynamic range, among others. Based on these parameters and the requirements of the user, different circuits are available. A voltage comparator could be used as an ADC as presented in Figures 2.21 and 2.22.

[†]Figure 2.19 was generated using the following Matlab code. $u=$ [-1:0.01:1]; $y=$uencode$(u,3)$; plot$(u,y,$‘.’$)$.

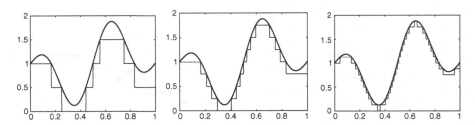

Figure 2.20 2, 4, and 8 bits. Higher resolution requires higher bit number

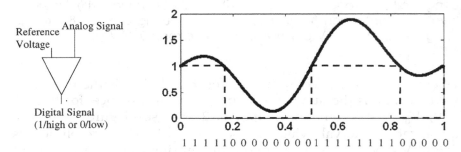

Figure 2.21 The output of a comparator is 1 bit high or low based on the reference

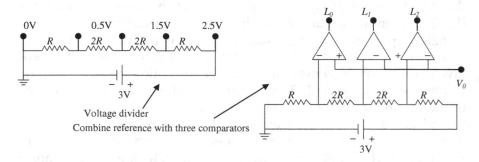

Figure 2.22 Combination of voltage dividers and comparators

---

†Figure 2.20 was generated using the following Matlab code. $i=[0:0.001:1]; u=\sin(\text{pi}*i).*\cos(3*\text{pi}*i)$; $n=4; y=u\text{encode}(u,n)$; $\text{plot}(i,\text{double}(y)/2^{\wedge}(n-1),'k')$; hold on; $\text{plot}(i,(u+1),'k.')$

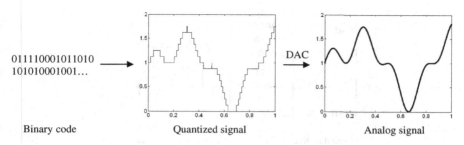

Binary code   Quantized signal   Analog signal

Figure 2.23 An example of DAC[†]

Flash ADC is one of the highest speed ADCs. This speed is limited mainly by the capacity of resistor and op-amp inputs. It does not need a clock but uses a lot of circuitry which drives the cost up. A 6 bit flash ADC requires 63 comparators. The accuracy of a flash ADC depends on the accuracy of the resistors and comparators used, and their reference voltage.

There is intrinsic error associated with the quantization or digitization process. For a uniform probability density function with round-off quantization error, and assuming that these errors are due to rounding and to negligible clipping, the signal-to-noise ratio is

$$\text{SNR} \approx (6n - 3.2)\text{dB}. \tag{2.3.5}$$

The commonly used 12-bit ADC SNR is about 68.8 dB. The SNR increases by 6 dB for each additional bit added to the ADC.

2.3.2 Digital-to-analog Converters

Digital-to-analog converters (DAC) are used to convert binary code into analog signals (see Figure 2.23) which are mainly used to transmit signals. The resolution or step size is the smallest change that can occur in the analog output as a result of a change in the digital input. It is defined as

$$\text{res} = A/(2^n - 1) \tag{2.3.6}$$

where A is the analog range and n is the number of bits. Different bits, which are weighted by a factor of two, affect the output voltage

[†]Figure 2.23 was generated using the following Matlab code. i=[0:0.0001:1];u=sin(2.3*pi*i).*cos (3*pi*i).^2;n=4;y=uencode(u,n); plot(i,double(y)/2^(n-1),'k');%hold on; plot(i,(u+1),'k'.)

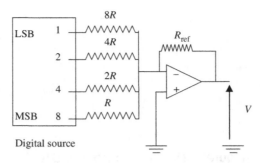

Figure 2.24 An example of a DAC where resistors are connected to the digital source

with different scale. Signal reconstruction involves holding the signal constant during the period between samples then passing the signal through a low-pass filter to remove high frequency components generated by the quantization process.

Full-scale output is the maximum analog output value. The circuit in Figure 2.24 is an example of a DAC.

2.3.3 ADC and DAC Errors

Noise is often a problem, especially in medium- to high-speed, high-resolution acquisition systems using high-speed processors. Figure 2.25, 2.26, and 2.27 illustrate some typical noise present in the ADC and DAC implementation that could be corrected for.

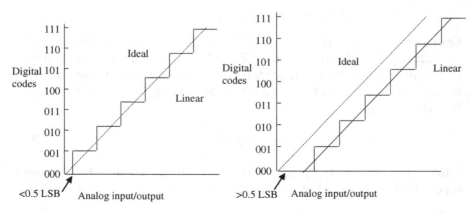

Figure 2.25 (a) Ideal ADC and DAC; (b) ADC and DAC with an offset error[15]

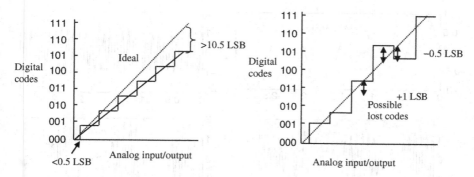

Figure 2.26 (a) ADC and DAC with a scale error; (b) ADC and DAC with non-monotic error with possible lost codes[15]

2.3.4 Modulation

Modulation is the process in which transmitted signal parameters vary with the message signal. This can give the signal a specific signature that could be isolated in an easier manner from the response or the echoed signal. This could also be within the sensor itself if the sensor has noise that needs to be distinguished from the measurement. Therefore coding or modulation could help in the error-detection and error-correction capabilities. Modulating a signal will increase the signal bandwidth which is one of the most important characteristics of any modulation technique. A narrow bandwidth is generally desirable since it allows more signals to be transmitted simultaneously than does a wider bandwidth. For

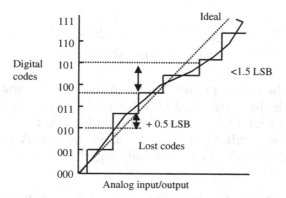

Figure 2.27 ADC and DAC with nonlinear error with lost codes[15]

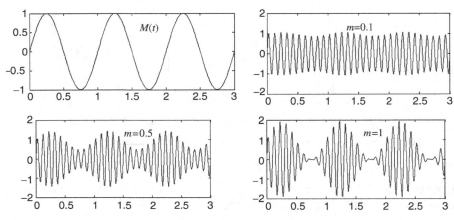

Figure 2.28 Modulation index ($m = 0.1$, 0.5, 1) effects on the message $M(t)$[†]

instance, a pure sine wave would exist at only one frequency therefore it will have zero bandwidth. However, a modulated signal is not necessarily a sine wave, and so it occupies a greater bandwidth.

Amplitude Modulation

In general, amplitude modulation (AM) is a technique in which the amplitude of the carrier is varied in accordance with some characteristic of the baseband modulating signal. It is the most common form of modulation because of the ease with which the baseband signal can be recovered from the transmitted signal. AM is performed using the equation:

$$s_{AM} = En \cdot V_C \sin(2\pi f_c t)$$
$$En = (1 + m \cdot M(t)) \tag{2.3.7}$$

where En is the envelope, $M(t)$ is the message (information) signal, $V_C \sin(2\pi f_C t)$ is the carrier and m is the modulation index. Figure 2.28 shows the impact of the modulation index on the message. The envelope is the resulting amplitude of the carrier. A product modulator is used to multiply two signals together.

[†]Figure 2.28 was generated using the following Matlab code. t=0:0.01:3;m=0.1;Env=1+m*sin$(2*3.14*t)$; sam=Env.*sin$(2*3.14*10*t)$;plot$(t$,sam);

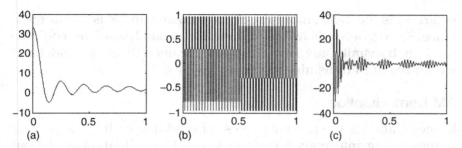

Figure 2.29 (a) is the message signal, (b) is the carrier signal and (c) is the transmitted AM modulated signal[†]

Two advantages of AM are its low bandwidth and its simplicity in modulating and demodulating a signal. However, AM presents some disadvantages such as the linearity of the message not being as conserved as other modulation schemes, and its vulnerability to interference with other signals as well as to noise for low-amplitude messages. For a single sine wave message of frequency f_m, then

$$s_{AM} = V_C(1 + m \cdot \sin(2\pi f_m t)) \sin(2\pi f_c t). \qquad (2.3.8)$$

Then analyze the frequency content of this AM signal using the trigonometric identity:

$$s_{AM} = V_C \sin(2\pi f_c t) + \frac{mV_C}{2}\cos(2\pi(f_c - f_m)t) - \frac{mV_C}{2}\cos(2\pi(f_c + f_m)t). \qquad (2.3.9)$$

The three terms represent, respectively, the carrier, the lower sideband (LSB), and the upper sideband (USB). Figure 2.29 illustrates a message, the carrier and the transmitted AM modulated signal.

Using equation (2.2.13) for the power of a sinusoidal voltage waveform, the carrier power P_c, where most of the power signal resides, and the power in either sideband P_{ISP} and P_{USP} for AM are

$$P_c = V_c^2/2R$$
$$P_{LSB} = P_{USB} = (mV_c/2)^2/2R = m^2 P_c/4 \qquad (2.3.10)$$

[†]Figure 2.29 was generated using the following Matlab code. For i=2:400; ti=(i-1)/400; xt(i)=ti; xt(1)=0;x1(1)=2+2*pi*5;x1(i)=2+sin(2*pi*5*ti)/(ti);end for i=1:400;ti=i/400;xt(i)=ti;x2 (i)= sin(2* pi*40*ti);end;x=x1.*x2;f=400*(0:256)/512;   subplot(131);plot(xt,x1);subplot(132);plot(xt,x2);subplot (133); plot(xt,x);

where V_c is the peak amplitude of the sinusoid and R is a load resistance. Since sidebands image one another, a bandpass filter could be used to transmit only one sideband. Transmitting one sideband would result in transmission power savings.

AM Demodulation

Demodulation is the reverse process of modulation. It is the process of message or information reconstruction. Demodulation of $x(t)$ out of $y(t)$ can be done by multiplying $y(t)$ with a locally generated oscillator that matches the carrier and passes it through a low-pass filter. There are two types of demodulation: synchronous and asynchronous. The synchronic demodulation is the case where there is no phase difference between the received signal and the carrier. Asynchronous demodulation is required when there is a phase difference. In order to recover the message from the total signal two conditions need to be satisfied. Firstly, the message should be positive, which can easily be done by adding a constant to $x(t)$ when modulating. Second, the carrier frequency should be larger than the highest frequency of the message. A low-pass filter is used to remove low frequencies. The message $x(t)$ can be recovered by an envelope detector as long as the two conditions are satisfied. Figure 2.30 illustrates a message modulation for a given modulation index, then demodulating it after mixing it with noise.

Frequency Modulation

The idea of frequency modulation (FM) is to change the frequency of a carrier signal with the amplitude of the input data signal. Frequency modulation is a technique in which the frequency of the carrier is varied in accordance with some characteristic of the baseband signal:

$$s_{FM}(t) = A_c \cos\left(k \int_{-\infty}^{t} m(r)dr + \omega_c t\right). \qquad (2.3.11)$$

The message is $m(t)$. The modulating signal is integrated because variations in the modulating term equate to variations in the carrier phase. The instantaneous angular frequency can be obtained by differentiating the instantaneous phase as shown

$$\omega = \frac{d}{dt}\left(k \int_{-\infty}^{t} m(r)dr + \omega_c t\right) = \omega_c + km(t). \qquad (2.3.12)$$

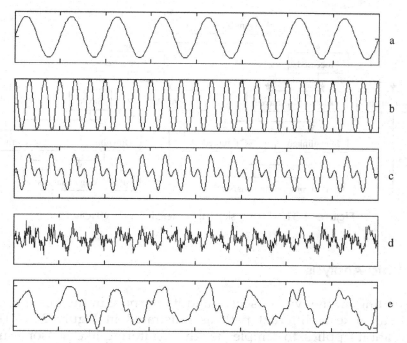

Figure 2.30 (a) is the message, (b) is the carrier, (c) is the signal (d) signal + noise, and (e) demodulated signal

Two of the advantages of FM modulation are constant transmitted power and resistance to noise. Transmitted power does not depend on $m(t)$ since it does not need to maintain strict amplitude linearity. Therefore it can use power efficient amplifiers. However, the drawback of FM is that, in addition to the complexity of the modulators and demodulators, FM requires more bandwidth than AM. Basically it trades bandwidth for SNR. Where Δf is the maximum deviation of the carrier frequency caused by the amplitude of the modulating signal, and f_m is the maximum frequency of the message signal, the modulation index becomes

$$m_f = \Delta f / f_m. \qquad (2.3.13)$$

FM signal bandwidth could be approximated by

$$B = 2(\Delta f + f_m). \qquad (2.3.14)$$

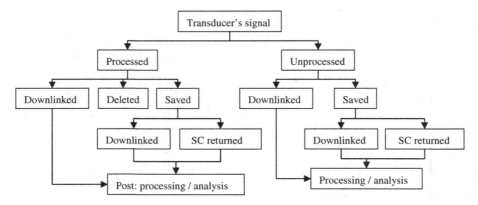

Figure 2.31 From signal collection to analysis

2.4 Data Analysis

Depending on the type of sensor and its purpose in the mission, the data can take a different path as illustrated in Figure 2.31. This illustration applies to simple health monitoring-like sensors such as temperature sensors and power sensors, as well as complex sensors such as radars.

Depending on resource allocation such as power, memory, data rate and coverage, the data is delivered to the user in one of the formats illustrated in Figure 2.31. Coverage here refers to the availability of a ground station to acquire the data. The downlink could be a direct downlink from the spacecraft to the ground station, or it could be routed to other spacecraft that have a direct ground access where users reside. Memory refers to on-board memory in any format. It could be in a solid state such as recording or magnetic tape. On the spacecraft, there are mainly two architectures that the data can follow. The distributed architecture, as shown in Figure 2.32, is where subsystem data are collected on the data bus. The traffic on

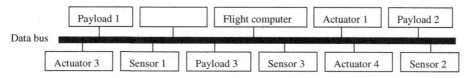

Figure 2.32 Distributed architecture

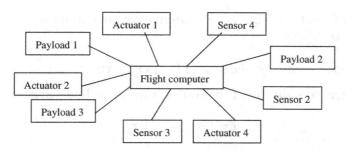

Figure 2.33 Centralized architecture

the data bus is typically regulated by a flight computer using standard communication protocols to exchange and route the data. The second architecture is the centralized or the star architecture, as shown in Figure 2.33, where each subsystem is directly connected to the flight computer. All data is generated by a sensor or, a subsystem.

Data processing can take place in various steps as illustrated in Figure 2.31. In this section few data analysis methods are presented. In general, regression analysis or curve fitting of data is widely used for the purpose of establishing a mathematical relationship between variables. This relationship could be used for calibration, forecast, prediction and validation of theoretical models. The data window depends on the processing time, and signal timing for post processing. The limiting factor would be how far into the data one may want to go. This approach could be used in any step of sensor measurement, such as calibration, sensor on, off, and warm-up.

In Figure 2.34 the distribution of y when x varies could be: (a) data seems to have a linear pattern, (b) data seems to have two linear patterns separated by a transition regime, (c) the data seem to follow a known non-linear pattern, and (d) there is no pattern between x and y and this means that the distribution is either random or

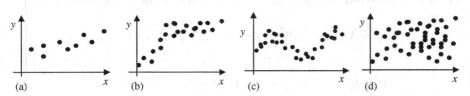

Figure 2.34 Typical two-dimensional data distributions

that a different variable or a combination of variables needs to be considered.

2.4.1 Uncertainty Analysis and Propagation of Error

If y is a real, continuous function of n independent variables:

$$y = f(x_1, x_2, \ldots, x_n), \tag{2.4.1}$$

then given small increments, Δx_1, Δx_2, ..., Δx_n, Δy the change in value of y, is

$$\Delta y = f(x_1 + \Delta x_1, x_2 + \Delta x_2, \ldots, x_n + \Delta x_n) - f(x_1, x_2, \ldots, x_n) = \Delta f. \tag{2.4.2}$$

The Taylor series expansion around x_1, x_2, ..., x_n is

$$\Delta f = \partial_{x_1} f \cdot \Delta x_1 + \partial_{x_2} f \cdot \Delta x_2 + \cdots + \partial x_n f \cdot \Delta x_n. \tag{2.4.3}$$

This is the change in value of a function f, given small changes in the values of its independent variables, x_i. In experimental uncertainty analysis, the Δs represent uncertainties in the estimates of physical quantities. Δs are equally likely to be of either sign, provided systematic biases have been removed or compensated for. A positive quantity can be measured by taking $(\Delta f)^2$. If the uncertainties are uncorrelated, independent variables then, on average, the most likely value of the uncertainty is

$$\left(\bar{\Delta f}\right)^2 = \left(\partial_{x_1} f \cdot \Delta x_1\right)^2 + \left(\partial_{x_2} f \cdot \Delta x_2\right)^2 + \cdots + \left(\partial x_n f \cdot \Delta x_n\right)^2. \tag{2.4.4}$$

Now compare two measurements in the case where one has the uncertainty in both measurements:

$$P = VI = RI^2 \tag{2.4.5}$$

For the case where measurements have the same uncertainty:

$$P = VI \Rightarrow (\Delta P)^2 = (\partial_V P \cdots \Delta V)^2 + (\partial_I P \cdots \Delta I)^2$$
$$\Rightarrow (\Delta P/P)^2 = (\Delta V/V)^2 + (\Delta I/I)^2 \tag{2.4.6}$$
$$P = RI^2 \Rightarrow (\Delta P)^2 = (\partial_R P \cdot \Delta R)^2 + (\partial_I P \cdot \Delta I)^2$$
$$\Rightarrow (\Delta P/P)^2 = (\Delta V/V)^2 + 4(\Delta I/I)^2. \tag{2.4.7}$$

This approach suggests that for lower uncertainty of the measurement, one should design the sensor based on measuring the voltage and the current. Measuring the resistance and current will increase the uncertainty by

$$3(\Delta I/I)^2. \tag{2.4.8}$$

2.4.2 Regression Analysis

Regression analysis is a formalized way to develop models or equations from historical data. It is a technique for curve fitting when the relationship between the dependent variables is not obvious. The approach is to assume an equation form, transform it to linear form if possible, perform least squares to fit the data and check how good the fit is. There are a number of standard equation forms that could be used such as those given in Table 2.3.

Polynomial regression is a regression that takes the form of

$$y = a_n x^n + a_{n-1} x^{n-1} + \cdots + a_1 x + a_0 \tag{2.4.9}$$

where n is the order of the polynomial regression. It is also possible to have a combination of regression within one set of data. This means that within the data-set, two or more domains could be identified where in each domain, a curve fit is obtained. Transient regimes are the areas where the two domains intersect. An acceptable regression analysis of the output of a sensor should have a high correlation coefficient. There would ideally be a mathematical relation between the variable and the output, searching for a pattern to express a phenomenon or to quantify a phenomenon. Regression analysis is

Table 2.3 Typical standard equation forms

Type	Equation
Linear	$y = ax + b$
Multiple linear	$y = a_0 + a_1 x_1 + a_2 x_2 + \cdots$
Hyperbolic	$y = 1/(ax + b)$
Polynomial	$y = a_0 + a_1 x + a_2 x^2 + \cdots$
Exponential	$y = ab^x \Rightarrow \log y = \log a + x \log b$
Geometric	$y = ax^b \Rightarrow \log y = \log a + b \log x$

used in order to identify the relationship between variables and phenomena that the sensor is measuring. It is a statistical tool that helps us to find this relationship for a set of data.

A simple linear regression between X and Y, which is the best line that fits the data, would be in the form of the equation $(y = ax + b)$. It is necessary to find the suitable a and b that best describe the data set. The vertical difference between an observed value and the line is called the residual or deviation error. Using the least squares method, the a and b coefficients that best fit the data can be determined when a scatter plot is used to establish a linear association. The next step is to obtain the best fitting regression line.

2.4.3 Least Square Method

The least square method is used to minimize the distance between the actual value and the estimated one. This estimated value could be derived from a prior knowledge of the measurement. It is derived via a mathematical equation or other means. The residual or error for a linear fit is

$$r_i = y_i - y_{i\_curve} = y_i - (a_1 x_i + a_0) \qquad (2.4.10)$$

where y_{i_curve} is the value on the curve of the ith data point. Since the residual is positive or negative, the square r_i^2 is summed. The goal will be to minimize this sum, which means that the derivative with respect to a and b of these sums are null.

$$\sum r^2 = \sum (y_i - a_1 x_i - a_0)^2$$
$$\partial_{a_0} \left(\sum r^2 \right) = 0 \qquad (2.4.11)$$
$$\partial_{a_1} \left(\sum r^2 \right) = 0.$$

Therefore,

$$a_0 = \bar{y} - a_1 \bar{x}$$
$$a_1 = \frac{\sum (x_i - \bar{x})(y_i - \bar{y})}{\sum (x_i - \bar{x})^2} = R \frac{\sigma_{xy}}{\sigma_{xx}} \qquad (2.4.12)$$

where R is the correlation coefficient. These are called the least squares values of a_0 and a_1. The x bar and y bar ($\bar{x}$ and $\bar{y}$)

are the averages of x and y, and σ_{xx} and σ_{yy} are the variances of x and y:

$$\sigma_{xx}^2 = \frac{1}{n-1}\sum_1^n (x_i - \bar{x})^2$$

$$\sigma_{yy}^2 = \frac{1}{n-1}\sum_1^n (y_i - \bar{y})^2. \tag{2.4.13}$$

For perfect correlation, $R = 1$, all points lie on the regression line. When $R = 0$ the line is useless. If the data is very scattered that means the wrong representative parameter was chosen. One has to work harder to either filter the data or just find a new form for the variable. Correlation coefficient R determines how good the curve fit is

$$R = \sqrt{1 - \sigma_{yx}^2 \Big/ \sigma_y^2}$$

$$\sigma_y^2 = \frac{1}{n-1}\sum_{i=1}^n (y_i - y_m)^2 \tag{2.4.14}$$

$$\sigma_{yx}^2 = \frac{1}{n-2}\sum_{i=1}^n (y_i - y_{i\_curve})^2$$

where σ_y is the standard deviation, y_i is the actual values of y, and y_{i_curve} is the value computed from the correlation equation for the same value of x. For an nth order polynomial fitting,

$$r_i = y_i - y_{i\_curve} = y_i - a_0 + a_1 x_i + \cdots + a_n x_i^n. \tag{2.4.15}$$

The least square method would be to minimize the quadratic function, and derive the coefficients a_i such as

$$r = \sum r_i^2 \Rightarrow \partial_{a_i} r = 0. \tag{2.4.16}$$

This approach is also generalized for any function, in that the approach is to minimize the quadratic sum error by taking the derivative of this quadratic error in respect to the fitting coefficients.

2.4.4 Fourier Analysis

A signal in the time domain can be represented in the frequency domain using the Fourier transform. There are many advantages in working in the frequency domain compared with the time domain. For instance, looking at the frequency content of a signal one can identify the main frequencies where most of the energy resides or other frequencies where the energy of the signal is wasted. Another advantage is that recording the frequency content often takes less space than recording the time domain of a signal. Losses in the recording can occur. For example a cosine function could be represented in the frequency domain by two delta functions. In general, signal synthesis is often conducted in the time domain and analysis in the frequency domain. The reverse process, where the signal is transformed from the frequency domain to the time domain, uses the inverse Fourier transform.

A function f could be represented as the sum of an odd function f_{odd} and an even function f_{even}

$$f(t) = f_{\text{odd}}(t) + f_{\text{even}}(t)$$

$$f_{\text{odd}}(t) = \frac{1}{2}(f(t) - f(-t)) \qquad (2.4.17)$$

$$f_{\text{even}}(t) = \frac{1}{2}(f(t) + f(t)).$$

Most odd and even functions can be written in the following way:

$$f_{\text{even}}(t) = \frac{1}{\pi}\sum_{0}^{\infty} a_k \cos(kt)$$

$$\qquad (2.4.18)$$

$$f_{\text{odd}}(t) = \frac{1}{\pi}\sum_{0}^{\infty} b_k \sin(kt)$$

where a_k and b_k are called the Fourier coefficients that define the Fourier series and represent the frequency content of the signal, and k is an integer. These coefficients can be determined by multiplying each function with the appropriate trigonometric function and integrating between $-\pi$ and $+\pi$

$$\int_{-\pi}^{+\pi} f_{\text{even}}(t)\cos(mt)dt = \frac{1}{\pi}\sum_{k=0}^{\infty} a_k \int_{-\pi}^{+\pi} \cos(kt)\cos(mt)dt \Rightarrow$$

$$a_k = \int_{-\pi}^{+\pi} f_{\text{even}}(t)\cos(kt)dt \qquad (2.4.19)$$

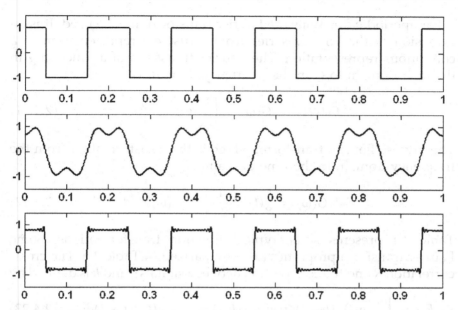

Figure 2.35 The square function, Fourier series for $n = 3$ and $n = 15$[†]

$$\int_{-\pi}^{+\pi} f_{\text{odd}}(t) \sin(mt)dt = \frac{1}{\pi} \sum_{k=0}^{\infty} b_k \int_{-\pi}^{+\pi} \sin(kt) \sin(mt)dt \Rightarrow b_k$$

$$= \int_{-\pi}^{+\pi} f_{\text{odd}}(t) \sin(kt)dt. \qquad (2.4.20)$$

For instance, using a square function, one can look at the first few terms of the Fourier series, and extend that to more terms of the Fourier series. Figure 2.35 shows the higher n is the better fit is the Fourier series in matching the square function. Once the fit is accepted, the frequency of the Fourier terms can be used to determine and identify this square function. These frequencies are the signature of the square function, and the Fourier coefficients are the weight of each frequency. The Fourier series converge at $1/k$ to the ideal function. Series are truncated when the error between the real value and the Fourier estimate is too small. This truncation results in saving data storage.

[†]Figure 2.35 was generated using the following Matlab code. $f=5$; for $k=0:15$; $j=2*k+1$;for $i=0:1000$; $t=i/1000$; $s(i+1)=(1/(j*(-1)^\wedge k))*\cos(2*\text{pi}*f*t*j)$; $x(i+1)=\text{square}(2*\text{pi}*f*t+\text{pi}/2)$;$st(i+1)=t$; end;$ss(1,:)=$ $0.*s$;$ss(k+2,:)=ss(k+1,:)+s$; end subplot (311);plot (st,x);subplot (312);plot $(st,ss(3,:))$;subplot (313);plot $(st,ss(15,:))$;$p=f*(j-1)$;

For aperiodic functions the Fourier transform may be used. It is an extension of the Fourier series from a discrete representation to a continuous representation. The Fourier transform of a function g in the time domain to G in the frequency domain is

$$F_T(g(t)) = G(\omega) = \int_{-\infty}^{\infty} g(t)e^{-j\omega t}dt. \tag{2.4.21}$$

The inverse Fourier transform, which is the transformation from the frequency domain to the time domain, is

$$F_T^{-1}(G(\omega)) = g(t) = \frac{1}{2\pi}\int_{-\infty}^{\infty} G(\omega)e^{j\omega t}d\omega. \tag{2.4.22}$$

Table 2.4 presents some typical function transformations. Some Fourier transform properties are summarized in Table 2.5. The cross-correlation R and the convolution between signals g and h is defined by

$$R_{gh} = \int_{-\infty}^{\infty} g(t)h(t-\tau)dt; g(t) \otimes h(t) = \int_{-\infty}^{\infty} g(t)h(\tau-t)dt. \tag{2.4.23}$$

The Fourier transform of the autocorrelation is

$$F_T(g(t) \otimes g(-t)) = G(\omega) \cdot G^*(\omega) = |G(\omega)|^2. \tag{2.4.24}$$

This is the power spectral density (PSD). Therefore the PSD of a signal is the Fourier transform of its autocorrelation. The signal energy could be computed in the time domain or in the frequency domain:

$$E = \int_{-\infty}^{+\infty} |g(t)|^2 dt = \frac{1}{2\pi}\int_{-\infty}^{+\infty} |G(\omega)|^2 d\omega = R_{gg}(0). \tag{2.4.25}$$

Table 2.4 Fourier transform and inverse Fourier transform of some popular functions

$g(t)$	$G(\omega)$
$\delta(t)$	1
$A, \infty \le t \le \infty$	$2\pi A\delta(\omega)$
$A\,rect(t/\tau)$	$A\tau sinc(\omega\tau/2)$
$\cos(\omega_o t)$	$\pi(\delta(\omega - \omega_o) + \delta(\omega + \omega_o))$
$e^{-at}u(t), a > 0$	$1/(a + j\omega)$
$e^{-a|t|}, a > 0$	$2a/(a^2 + \omega^2)$
$e^{j\omega_o t}$	$2\pi\delta(\omega - \omega_o)$
$\sin(\omega_o t)$	$j\pi(\delta(\omega + \omega_o) - \delta(\omega - \omega_o))$

Table 2.5 Fourier properties where lower case letters are in the time domain and upper case letters are in the frequency domain or Fourier space

Fourier property	Time domain	Frequency domain
Linearity	$af(t) + bg(t)$	$aF(\omega) + bG(\omega)$
Time-shifting	$f(t - t_0)$	$e^{-j\omega t_0} F(\omega)$
Time-reversal	$g(-t)$	$g(-t) \leftrightarrow g(-\omega)$
Time-scaling	$g(at)$	$G(\omega/a)/|a|$
Differentiation	$dg(t)/dt$	$j\omega G(\omega)$
Integration	$\int_{-\infty}^{t} g(t)dt$	$\dfrac{1}{j\omega} G(\omega) + \pi G(0)\delta(\omega)$
Frequency shifting	$e^{j\omega_0 t} g(t)$	$G(\omega - \omega_0)$
Differentiation in frequency	$tg(t)$	$jdG(\omega)/d\omega$
Symmetry	$g(t)$	$2\pi G(-\omega)$
Multiplication	$g(t) \cdot h(t)$	$G(\omega) \otimes H(\omega)$
Convolution	$g(t) \otimes h(t)$	$G(\omega) \cdot H(\omega)$

To find the PSD of a random signal, the correlation is often used. The convolution is used typically to determine the output of linear systems.

Bibliography

1. Baher, H. (2001) *Analog and Digital Signal Processing*, John Wiley and Sons.
2. Bendat, J. S. and Piersol, A. G. (1971) *Random Data: Analysis and Measurement Procedures*, Wiley-Interscience.
3. Blake, R. (2002) *Electronic Communication Systems*, Delmar.
4. Brown, R. G. (1983) *Introduction to Random Signal Analysis and Kalman Filtering*, John Wiley and Sons.
5. Fraden, J. (1997) *Handbook of Modern Sensors*, American Institute of Physics.
6. Holman, J. P. (1994) *Experimental Methods for Engineers*, McGraw-Hill, Inc.
7. Jones, N. B. and Watson, J. D. M. (1990) *Digital Signal Processing: Principles, Designs and Applications*, Peter Peregrinus Ltd.
8. Larson, W. J. and Wertz, J. R. (1992) *Space Mission Analysis and Design*, Microcosm, Inc.
9. Mohanty, N. (1991) *Space Communication and Nuclear Scintillation*, Van Nostrand Reinhold.
10. NASA (2001) *Dynamic Environmental Criteria*, NASA-HDBK 7005.

11. Schwartz, M. and Shaw, L. (1975) *Signal Processing: Discrete Spectral Analysis, Detection and Estimation*, McGraw-Hill.
12. Sloggett, D. R. (1989) *Satellite Data Processing, Archiving and Dissemination*, Ellis Horwood Limited.
13. Taylor, R. H. (1997) *Data Acquisition from Sensor Systems*, Chapman and Hall.
14. Willis, J., Tompkins, W. J. and Webster, J. G. (1987) *Interfacing Sensors to the IBM PC*, Prentice Hall.
15. Young, T. (1985) *Linear Systems and Digital Signal Processing*, Prentice-Hall, Inc.

3

Noise and Filtering in Spacecraft Sensors

It is well established that any signal measurement is associated with a noise source, which could limit the accuracy and the precision of a measurement. This noise source is sometimes known and characterized, sometimes eliminated and at other times it is just an intrinsic part of the measurement. The challenge is to figure out how to improve the signal while minimizing the noise. There are limits set by the nature of the phenomena of interest, but the process of analysis, measurement, proximity and the level of understanding can eventually reach these limits. Eliminating noise in spacebased applications could be limited by cost. For example, shielding is a typical approach to reduce radiation; however, adding shielding is often associated with mass increase which is directly related to cost. There are many reasons to use spacebased sensors. For example, spacebased sensors offer the capability of wide coverage in real time. Spacebased sensors could be used as a substitute for groundbased sensors in order to minimize noise. For instance, Hubble is a spacebased telescope that does not face the same atmospheric perturbation that earthbased telescopes face. Earth's atmosphere is considered as noise for astronomical observation. One should, however, keep in mind that noise is also relative to the measurement of interest, in that where one measurement considers noise as a handicap, other measurements are actually composed of the noise itself. In other words, what is noise for one measurement is sometimes the actual signal of

Spacecraft Sensors.   Mohamed M. Abid
© 2005 John Wiley & Sons, Ltd

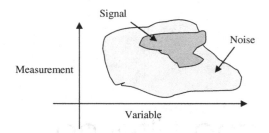

Figure 3.1 Nature marries signals with noise

interest for another. The following figure shows a typical measurement of a variable where the signal is embedded in the noise. Sometimes changing the measured variable would cause the signal to be much larger and the noise much smaller. A clear understanding of the phenomenon of interest would help to reshape the graph in Figure 3.1.

In this chapter, various sources and types of noise that are relevant to spacecraft sensors are identified and quantified, typical approaches to minimize noise are presented and filters that reduce some types of noise are introduced.

3.1 Internal Noise

3.1.1 Thermal Noise

Thermal noise, also known as Johnson noise, is a noise that results from the fluctuation caused by thermal motion of the charge carrier in a resistive element. Electrons moving randomly under the influence of temperature generate fluctuating voltage. Power can nevertheless be delivered to another device at a lower temperature. The thermal noise power P_{ThN} is constant with frequency. There is equal power in every hertz of bandwidth. It is sometimes called white noise as an analogy to white light that is a blend of all colors.

$$P_{ThN} = kT\beta$$
$$P_{ThNdB} = 10 \log kT\beta$$

(3.1.1)

where P_{ThNdB} is thermal noise power in dB, k is the Boltzmann's constant 1.38×10^{-23} JK^{-1}, T is the absolute temperature of the system (in K), and β is the receiver noise bandwidth being considered (in Hz).

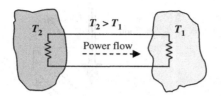

Figure 3.2 Thermal noise

A bandwidth reduction will imply a thermal noise reduction. For a temperature of 290 K, and a bandwidth of 1 Hz, $P_{ndB} = -204$ dBW $= -174$ dBm. For a bandwith of 1 MHz, $P_{ndB} = -114$ dBm. This equation demonstrates that the thermal noise can be reduced by cooling the output amplifier or by decreasing its electronic bandwidth, both of which impact the measurement. One has to trade-off between low noise performance and speed of readout. Voltage fluctuations are greater across a hotter resistor and the power is transferred from the hotter resistor to the colder resistor until eventually thermodynamics equilibrium is reached when $T_2 = T_1$ (see Figure 3.2).

Any thermodynamic system stores (at equilibrium) $\frac{1}{2}kT$ energy per degree of freedom D_{oF}

$$V_n^2/R = kT\beta D_{oF}/2 \Rightarrow V_n = \sqrt{kT\beta D_{oF}R/2} \qquad (3.1.2)$$

where D_{oF} is the degree of freedom per unit bandwidth. There are two D_{oF} for two directions of energy transfer or for electric plus magnetic fields or for an arbitrary polarization direction

$$|V_J| = \sqrt{4kT\beta R} \qquad (3.1.3)$$

where V_J is the RMS noise voltage and R is the resistance of the circuit element. In most systems, the minimum frequency f_{min} would be close to zero (DC) so the bandwidth is the maximum frequency f_{max}. At room temperature the peak-to-peak voltage is given by the equation:

$$|V_{peak-to-peak}| = 6.51 \times 10^{-10} \sqrt{fR}. \qquad (3.1.4)$$

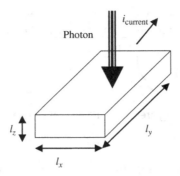

Figure 3.3 Photoconductor illumination[5]

The magnitude of thermal noise is generally in the μV range and can often be ignored. For optic based detectors, such as photoconductors, the detector resistance R_d is

$$R_d = \rho l_y / l_x l_z \qquad (3.1.5)$$

where ρ is the detector material resistivity, l_y is the absorption length, $l_x l_z$ is the optical active area A_d for light. Figure 3.3 illustrates a typical illumination of a photoconductor.

3.1.2 Thermal Emf

Temperature gradients in conductors of different materials can induce small thermoelectric voltages by the Peltier effect. The error that results from this effect in resistors is often overseen. Thermocouples use the Peltier effect to measure temperature. The Peltier effect could be used as a heater or as a source of power. To minimize the Peltier voltage one can keep same metals in positive and negative leads, or keep leads together so that they are exposed to the same temperatures or minimize temperature gradients within a measuring device.

3.1.3 Parameter Noise

The parameter being measured may be noisy. For example, using a GPS to measure a spacecraft's altitude but placing it on the very end of the extent of the solar panel can cause the returned signal to be noisy due to the solar panel dynamics. It is crucial to understand the parameter that is to be measured, to be aware of the adjacent instruments and then to properly accommodate the sensor on the spececraft.

3.1.4 Dark Current

In a photovoltaic device electrons can be generated either by thermal motion of the silicon atoms or by the absorption of photons. Electrons produced by these two effects are indistinguishable. Thermal motion of atoms causes a current – a dark current. Dark current is analogous to the fogging that can occur with photographic emulsion if the camera leaks light. It can be reduced or eliminated entirely by cooling the sensing element. Dark current is governed by Poisson statistics. For instance, if the mean dark current in an image is 400 electrons per pixel, the noise contribution into the measurement of any photo-charge pixel is 20 electrons.

3.1.5 Shot Noise

An electric current is composed of a stream of discrete electrons. When the current is small, or the sampling time is short, a typical sample may contain a relatively small number of electrons, less than about 10^6, for example. Statistical fluctuations from sample-to-sample constitute shot noise. Shot noise has a simpler power spectrum than thermal noise. It results from the random fluctuations in a DC current flow across a potential barrier. The noise on these electrons follows the Poisson equation. Poisson fluctuations of charge carrier number, such as the arrival of charges (see Figure 3.4), induce charges on electrodes which are quantized in amplitude and time.

Photon noise, also called shot noise, occurs when photons arrive in an unpredictable fashion described by Poisson statistics. This unpredictability causes noise. Using Poisson statistics, the root mean square (RMS) uncertainty noise in the number of photons per second detected by a sensing element is equal to the square root of the mean photon flux, which is the average number of photons detected per

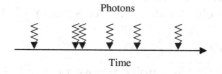

Photons

Time

Figure 3.4 Time of photon arrival on a sensing element

second. For instance, for an imaged star where a pixel reads an average N photo-electrons per second, then the brightness uncertainty of the measurement would be the square root of N. Increasing the exposure time T would increase the photon noise. It will also increase the signal-to-noise ratio (SNR), which is highly desirable. The signal-to-noise ratio increases as the square root of T if the only noise present is the photon noise.

For an ideal photocathode, the current flow is

$$i = n_e q / \tau \qquad (3.1.6)$$

where n_e is the number of photo-generated electrons in the time interval τ, and q is the electron charge -1.6×10^{-19} C. For very small τ, where either one electron is present or none in that time interval, a statistical binomial distribution is used. For a large number of electrons, the distribution follows Poisson statistics. The average current i_{av} is related to the average number of electrons created n_{av}:

$$i_{av} = n_{av} q / \tau. \qquad (3.1.7)$$

The variance of n_e equals the mean number of electrons if Poisson statistics are assumed:

$$\left. \begin{aligned} i_s^2 = \overline{(i - i_{av})^2} = \frac{q^2}{\tau^2} \overline{(n_e - n_{av})^2} = \frac{q^2}{\tau^2} n_{av} \\ \Delta f = \frac{1}{2\tau} \end{aligned} \right\} \Rightarrow i_s = \sqrt{2q i_{av} \Delta f}. \qquad (3.1.8)$$

This noise is proportional to the square root of the detector area if the potential barrier cross-sectional area is an optically active area, which is typical for a photodiode. The average current is

$$i_{av} = P_{signal}\, \rho_r G^2 \Rightarrow i_s^2 = 2q P_{signal}\, \rho_r \Delta f \cdot G^2 \qquad (3.1.9)$$

where ρ_r is the current responsivity in A/W, and P_{signal} is the power signal.[5] The responsivity corresponds to the detector's conversion of optical power into electric current (see section 4.5.3).

3.1.6 Excess Noise or 1/f Noise

Flicker or excess noise is a noise whose power is frequency dependent, sometimes called $1/f$ noise. It is more important for lower frequencies. It is also called pink noise because there is more energy at the lower frequency end of the spectrum than with white noise. It is found, for example, in electron tubes and it is commonly associated with interface states in MOS electronics. Unfortunately, it is not yet fully understood.

3.1.7 Dielectric Absorption

When measuring very small currents at the pico-amp level, the charge storage effects of the insulators in the system can become significant. This effect only occurs with high impedance ($10^6 \Omega$) sensors and only when the sensor output changes relatively quickly. Dielectic absorption is more of a signal distortion than superimposed noise. However, the effect can amplify the impact of other noise sources. Sensors that exhibit sensitivity to dielectric absorption noise are also likely to be susceptible to cable noise. To minimize this noise, low absorption insulators can be used to minimize the magnitude of voltage changes to less than a few volts.

3.2 External Noise

3.2.1 Clutter Noise

There are varieties of clutter noise or background sources that a detector is subject to. For instance, solar radiation backscattered by surrounding objects can contaminate a radiometric measurement of a target that is embedded within the surrounding objects. This background radiation could be minimized by implementing a narrow-band filter in front of the sensing element, for example. For an optical receiver the background power is

$$P_b = N_\lambda A \Omega \Delta\lambda \eta_f \eta_o \eta_a \qquad (3.2.1)$$

where N_λ is the spectral radiance at the receiving aperture in $Wm^{-2}\mu m^{-1}sr^{-1}$, A is the receiver aperture area in m^2, Ω is the receiver acceptance solid angle in rad^2, $\Delta\lambda$ is the spectral filter band-pass in μm, η_f is the target area fill factor in the receiver field of view, η_o is the receiver optical efficiency, and η_a is the atmospheric transmission. There is also the backscattered radiation from the intermediate environment that separates the target and the sensor, mainly observed in laser-based sensors. The clutter power due to backscatter depends on the pulse length which means that using short pulses can reduce this clutter noise.[14]

$$P_c = P_T A \beta g \eta_0 \eta_a \frac{ct_p}{2z^2}$$

$$g = \begin{cases} 1; & \Omega \geq \Omega_0 \\ \pi\theta^2/\Omega_0; & \Omega < \Omega_0 \end{cases} \tag{3.2.2}$$

where P_c is the clutter power due to backscatter from suspended particles along the propagation path, t_p is the pulse width in seconds, β is the intermediate backscatter coefficient in $m^{-1}sr^{-1}$, z is the range in m, Ω is the solid-angle field of view of the receiver, Ω_0 is the solid angle subtended by the transmitter beam, and $c = 3 \times 10^8$ m/s.

3.2.2 Jamming

Jamming is used intentionally in electronic warfare to impact the performances of the detection equipment such as radars, communication systems and GPS receivers. A jammer could be in the form of a transmitter that emits RF signals tuned to frequencies that include the one of the receiver. The transmitted signal would also have a similar modulation to the receiver. A jammer could receive the signal, alter it and then retransmit the modified signal back to the radar. This can saturate a receiver with a false signal. It can delay or deny target detection. There are many types of jamming as illustrated in Figure 3.5.

Broadband jamming occurs over a broad band of frequencies. Noise or random signals are emitted over a large bandwidth which limits the jamming strength (see Figure 3.5). Partial-band jamming occurs when the output power of the jammer is centered on a narrow

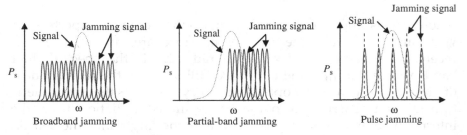

Figure 3.5 Typical types of jamming

bandwidth that overlaps partially with the known frequency of the receiver. Pulse jamming is a jamming technique that uses high power pulses which are emitted at a very narrow frequency band. This type of jamming could be used to interfere with an unknown frequency within a known range.

Factors which affect the vulnerability to jamming are the antenna patterns of receiving antennas, link margin, signal bandwidth and data redundancy. Encryption and modulation could minimize the impact of jamming on the sensor.

3.2.3 Radio Frequency Coupling

This type of signal is not trivial to assess or to model. Circuits or cables can act as an antenna that receives external electromagnetic waves. This could result in a fluctuating DC input voltage. Using shielded cable and grounding the shield to the case of the measuring device could resolve this problem. Capacitive coupling the end of the shield could avoid ground loops. One can also place ferrite beads over the signal wires including signal grounds, at both ends of the sensor cable.

3.2.4 Electromagnetic Field Coupling

Varying electric fields can be coupled into a measuring circuit by the stray capacitance between the two circuits. This capacitance is usually very small, of the order of a few pF, but it is sufficient to introduce voltage noise into high impedance circuits. Shielded cable connected to the ground at one end, increasing the distance between the electric field source and the measuring circuit and lowering the measuring circuit impedance as much as accuracy considerations allow are some typical approaches to minimize this noise. Also noise could be induced in a pair of wires from nearby wire pairs. Commonly, this occurs when wires are placed closed together. The cumulative effect of noise is equal on both sides. This problem is remedied by twisting the pairs together, hence the term twisted pair. Twisting does not always eliminate the noise, but it does significantly reduce crosstalk interference which is the interference of the signal of adjacent pairs. It is most severe at the ends where the wires are untwisted. This is referred to as terminal crosstalk interference.

A twisted pair consists of two insulated copper wires arranged in a regular spiral pattern to minimize the electromagnetic interference between adjacent pairs. There are two types of twisted pair: the unshielded twisted-pair UTP cable, which consists of two conductors each surrounded by an insulating material, and the shielded twisted-pair STP cable, which is a UTP cable with metal casing such as foil or braided-mesh that prevents electromagnetic noise. It also eliminates crosstalk, which occurs when one line picks up the signal of another line.

3.2.5 Inductive Coupling

A variation in magnetic fields induces a current into a measuring circuit. This current tends to flow regardless of the circuit impedance. The magnitude of the induced current is proportional to the magnetic field strength, the rate at which it changes and the area of pickup loop in the measuring circuit. One of the following approaches could be used to minimize this effect – reducing the circuit area, using twisted pair cables, mu-metal shielding, placing the measurement circuit away from magnetic fields, avoiding movement or vibration of magnetic materials near measurement circuits and tying down cables to prevent vibration.

3.3 Signal-to-Noise Ratio

Noise power or voltage is crucial to investigate and quantify. This allows the signal of interest to be isolated and analyzed since it is directly related to the performance of the sensor. What really matters is the relative amount of noise compared to the signal level. The signal-to-noise ratio (SNR) gauges the ability to isolate the signal from the noise. The SNR is typically expressed in decibels

$$SNR = P_s/P_n$$
$$SNR_{dB} = 10 \log(P_s/P_n) \tag{3.3.1}$$

where P_s is the signal power and P_n is the noise power. Ideally the *SNR* is infinite, which means that there is virtually zero noise. This is not representative of a real system. *SNR* can be difficult to measure. It is common to find in the specifications the ratio $(S+N)/N$ rather than S/N, which is the ratio of the signal with noise (measured when the sensor or transducer is on) to the noise when it is off.

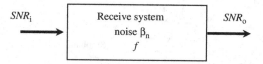

Figure 3.6 The transfer function of the receiver is the noise factor

Another parameter that is often used is the noise factor f which is the ratio of the input signal-to-noise ratio to the output signal-to-noise ratio from a sensor:

$$f = \frac{(S/N)_{\text{input}}}{(S/N)_{\text{ouput}}}. \tag{3.3.2}$$

The noise figure N_F is the noise factor expressed in dB

$$N_F = 10 \log(f). \tag{3.3.3}$$

Figure 3.6 shows a typical receiver system where the SNR_{out} depends on the SNR_{in}. The total output noise N of a receiver is proportional to the noise power P_n of an ideal receiver:

$$N = f P_n. \tag{3.3.4}$$

For the case where the noise is a thermal-like noise, or Johnson noise, which results from the thermal motion of the electrons in the resistive portion of the receiver transducer, or antenna, the noise interferes with the detected signal that is derived from the target, and the noise power is a function of the temperature T_{sys}. Equation (3.1.1) becomes

$$N = f P_n N_F = k T_{\text{sys}} \beta_n \cdot f \Rightarrow N_{\text{dB}} = 10 \log_{10}(k T_{\text{sys}} \beta_n) + N_F \tag{3.3.5}$$

where N_{dB} is the noise power in dBW that is relative to 1 W.

3.4 Filter Types

There are many ways to classify filters. They can be classified by the bandwidth or by the fraction bandwidth. For instance, narrow bandwidth filters have a typical relative bandwidth of less than

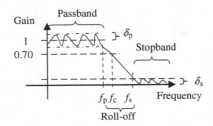

Figure 3.7 Filter specification – δ_p passband deviation, δ_s stopband deviation

5%, moderate bandwidth is between 5 and 25%, and wide bandwidth is larger than 25%. Filters can also be classified based on their transmission medium, such as microstrip, coaxial lines, lumped elements or waveguides. Ultimately, the frequency of operation would weight it one way or another. Filter specification includes performance specification (see Figure 3.7). Environmental performance of a filter is also a way of classifying filters, such as temperature range of operation and cycling. Size, shape and mounting surfaces would classify a filter based on mechanical specification.

Filters are frequency selective devices that pass certain frequencies of an input signal and attenuate others. Typical filters are low-pass, high-pass, band-pass and notch filters. A filter is classified by certain characteristics. These include frequency characteristics, such as the cut-off frequency f_c, the passband frequency f_p, often called bandwidth, and the stopband frequency f_s. The passband is the range of frequencies that pass through the filter. The cut-off frequency f_c is typically the frequency at which the gain of the filter drops by 3 dB from its maximum value in the passband. The stopband is similar to the passband but located in the frequency regions where the amplitude is minimal. The stopband frequency f_s locates the range of frequencies to be attenuated. Other classifying characteristics include the roll-off or transition between the stopband and the passband, the peak ripple in the passband δ_p, and the peak ripple value in the stopband δ_s. The gain-roll-off is the transition between the passband and the stopband. It is the region between f_c and f_s (see Figure 3.7). The closer f_c is to f_s the steeper is this transition region. The rate of this roll-off is determined by the order of the filter. The ratio $A_s = f_s/f_c$ is called the steepness factor and it is used for filter normalization.

Practical specifications are often given in terms of a loss function in dB:

$$G(\omega) = -20 \log_{10} |G(e^{j\omega})| \qquad (3.4.1)$$

peak passband ripple:

$$\alpha_p = -20 \log_{10}(1 - \delta_p) \qquad (3.4.2)$$

and minimum stopband attenuation:

$$\alpha_s = -20 \log_{10}(\delta_s). \qquad (3.4.3)$$

For digital filter design, normalized band-edge frequencies are typically computed from specifications in Hz using

$$\begin{aligned} \omega_{np} &= 2\pi f_p / f_t \\ \omega_{ns} &= 2\pi f_s / f_t \end{aligned} \qquad (3.4.4)$$

Each filter creates a systematic delay in the signal called group delay. It is the delay seen by all signal frequencies as they pass through the filter. This delay depends on the nature or the characteristic of the filter. It could be a constant delay to all frequencies within the passband such as Bessel filters, or it could be in the form of a rate of change in the phase with frequency seen in Butterworth, Chebyshev, inverse Chebyshev and Cauer filters. For these filters the group delay increases with frequency and reaches a peak value by the cut-off point, then gradually reduces to a constant value. The group delay increases with the number of stages or the order of the filter.

The filter transfer function is

$$H(f) = \frac{V_2(f)}{V_1(f)} = \frac{a_0 + a_1 s + \cdots + a_m s^m}{b_0 + b_1 s + \cdots + b_n s^n} \qquad (3.4.5)$$

$$s = i2\pi f = i\omega; \; A_V(f) = 20 \log_{10}(|H(f)|)$$

where n is the order of the filter. The filter response is determined by the polynomial coefficients. Popular polynomials have specific filter names such as Butterworth, Chebyshev, Elliptical and Bessel.[25]

The design or the choice of the filter is determined by the requirement or the specification of the system. Specifications such as the shape of the passband, wrinkled or not, attenuated or not, fast roll-off

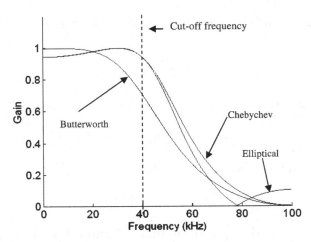

Figure 3.8 Three low-pass filters type behaviour for a cut-off frequency of 40 KHz[†]

and the extent of the group delay can affect the design. In practice the magnitude response specifications of a digital filter in the passband and in the stopband are given with some acceptable tolerances. Figure 3.8 illustrates some of these filters.

The Bessel response is smooth in the passband, and attenuation increases smoothly in the stopband. The stopband attenuation increases very slowly until the signal frequency is several times higher than the cut-off point. It has a very slow change of attenuation beyond the passband, but it has a great impulse performance. There is, however, poor frequency selectivity compared to other response types.

The Butterworth response has a smooth passband and a smooth increase in the stopband attenuation. It differs from the Bessel response by the attenuation in the stopband which rises by $n \times$ 6dB/octave almost immediately outside the passband. Typically the transfer function is all-pole with roots on the unit circle

Chebyshev filters have ripples in the passband. This allows the initial rate of the attenuation to increase more rapidly with frequency than a Butterworth filter of equal order. The stopband smoothly increases attenuation. The Chebyshev filter has faster roll-off. Because of the ripples in the passband, the stopband attenuation rises sharply just by the cut-off frequency f_c. Attenuation beyond f_c

[†]Figure 3.8 was generated using the following Matlab code. $[b,a]$=butter(2,0.4);%$[b,a]$=cheby1(2, 0.5,0.4);$[b,a]$=ellip(2,0.5,20,0.4);$[h,w]$=freqz(b,a,1024);freq=(0:1023)/(0.01*1024);hold on;plot(freq, abs(h));xlabel('Frequency (kHz)');ylabel('Gain').

rises by $n \times 6$ dB/octave similar to the Butterworth. However, the Chebyshev produces more stopband attenuation. The disadvantage of the Chebyshev filters is in the time domain, since its group delay is longer near the passband edge than the Butterworth response in addition to the ripples in the group delay. Typically the transfer function is all-pole with roots on an ellipse which results in a series of ripples in the passband.

Cauer filters, or elliptical filters, have ripples in the passband and in the stopband and a very sharp or steep roll-off due to the zeros of the transfer function. The circuit is, however, more complex and the group delay is larger. Inversely, the Chebyshev filter response has a smooth passband and nulls in the stopband. This leads to a sharp roll-off in the frequency response and a moderate overshoot in the impulse of a Cauer filter. The advantage of this filter is that the Q factor of its components, meaning the voltage or the current magnification, is lower than in the Chebyshev design. However, they are more complex and require more components.

In the following sections, low-pass, high-pass, band-pass and notch filters are introduced in both the passive and active modes. Passive filters contain only passive elements such as resistors, capacitors and inductors. The advantage of a passive filter, such as an LC filter, is that for frequencies above a few hundred kHz they are not inherently noisy. Disadvantages of passive filters include the fact that they require precision capacitors and inductors and are difficult to tune automatically, in addition to the large number of stages needed to get high order (typically $n = 2$ per stage). Active filters, which are based on op-amps, are good for frequencies up to a few hundred kHz. They exhibit low noise and interference if set up properly and they still require precision capacitors and resistors. Switched capacitor filters such as integrated circuits are good for frequencies up to a few tens of kHz. There are no precision components required and they are tuned by adjusting the clock frequency which is easy to tune automatically. They can get $n = 4$ per stage, two filters per IC. However, they have relatively poor DC gain control, and they require a simple (first order) pre-filter.

3.4.1 Low-pass Filter

A low-pass filter passes signals below its cut-off frequency ω_c and attenuates or rejects those above. Figure 3.9 shows a typical low-pass

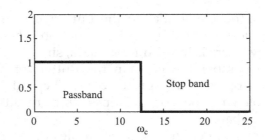

Figure 3.9 Ideal low-pass filter

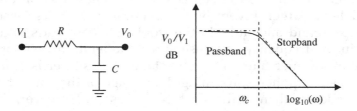

Figure 3.10 Passive first-order low-pass filter

filter. Figure 3.10 illustrates a passive first-order RC low-pass filter implementation

$$\frac{V_{\text{out}}}{V_{\text{in}}} = H = \frac{1/RC}{j\omega + 1/RC}$$

$$H(j\omega) = \frac{1}{1 + j\omega\tau} \tag{3.4.6}$$

$$\omega_c = \frac{1}{RC}; \tau = RC$$

In active filters an op-amp is added as well as a feedback loop. This loop, for a low-pass filter, is composed of capacitors (see Figure 3.11).

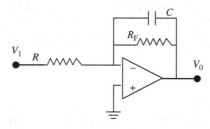

Figure 3.11 Active first-order low-pass filter

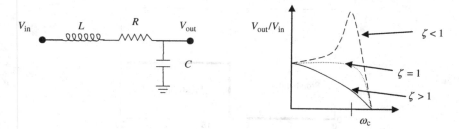

Figure 3.12   Passive second-order low-pass filter

The frequency response of this circuit is identical to the passive filter, but with the advantage of negligible loading because of the low output impedance of the op-amp

$$\frac{V_{out}}{V_{in}} = H(j\omega) = K\frac{\omega_c}{j\omega + \omega_c}$$

$$K = 1 + \frac{R_F}{R_1}; \ \omega_c = \frac{1}{RC}.$$

$$(3.4.7)$$

To increase the attenuation of the transfer function, the order of the filter could be increased by adding a certain number of capacitors and inductors. Figure 3.12 shows a passive second-order low-pass filter:

$$\frac{V_{out}}{V_{in}} = H(j\omega) = \frac{1}{(j\omega/\omega_c)^2 + (2\xi j\omega/\omega_c) + 1}$$

$$\zeta = \frac{R}{2}\sqrt{\frac{C}{L}}; \ \omega_c = \sqrt{\frac{1}{CL}}.$$

$$(3.4.8)$$

Depending on ζ, the response is under-damped when $\zeta < 1$, critically damped when $\zeta = 1$, or over-damped when $\zeta > 1$. Similarly, the order of low-pass active filters could be increased. Figure 3.13 shows a higher-order active low-pass filter.

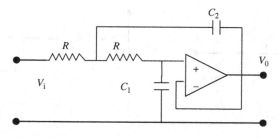

Figure 3.13   Active high-order low-pass filters

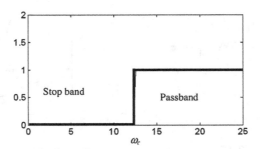

Figure 3.14 Ideal high-pass filter

3.4.2 High-pass Filter

High-pass filters reject signals below their cut-off frequency and pass those above (see Figure 3.14). Figure 3.15 illustrates a passive first-order RC high-pass filter implementation. Switching R and C from the low-pass filter (Figure 3.10) one gets a high-pass filter as shown in Figure 3.15:

$$\frac{V_{out}}{V_{in}} = H = \frac{R}{Z} = \frac{s}{s + 1/RC}$$

$$H(j\omega) = \frac{j\omega\tau}{1 + j\omega\tau} \quad (3.4.9)$$

$$\omega_c = \frac{1}{\tau}; \ \tau = RC; \ s = j\omega.$$

Figure 3.16 shows a typical active first-order high-pass filter:

$$\frac{V_{out}}{V_{in}} = H(s) = K\frac{s}{s + 1/RC}; \ K = 1 + \frac{R_F}{R_1}; \ \omega_{co} = \frac{1}{RC}; \ H(s) = K\frac{s}{s + \omega_{co}}.$$

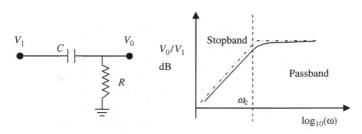

Figure 3.15 First-order high-pass filter

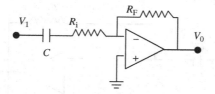

Figure 3.16　Active first-order high-pass filter

For normalized components of a low-pass filter, the capacitors of low-pass filters can be changed to inductors (see Figure 3.12) and the inductors can be changed to capacitors in order to form a high-pass filter (see Figure 3.17):

$$C_{\text{low-pass}} = 1/(\omega_c L_{\text{high-pass}})$$
$$L_{\text{low-pass}} = 1/(\omega_c C_{\text{high-pass}}) \tag{3.4.10}$$

where C and L are, respectively, normalized capacitance and inductance, and ω_c is the cut-off frequency. To increase the attenuation of the transfer function, the order could be increased by adding a capacitor and an inductor to make a second-order high-pass filter.

$$\frac{V_{\text{out}}}{V_{\text{in}}} = H(j\omega) = \frac{\omega^2}{(j\omega/\omega_c)^2 + (2\zeta j\omega/\omega_c) + 1}$$
$$\zeta = \frac{R}{2}\sqrt{\frac{C}{L}}; \; \omega_c = \sqrt{\frac{1}{CL}} \tag{3.4.11}$$

Similarly to previous filters, depending on ζ, the response is under-damped when $\zeta < 1$, critically damped when $\zeta = 1$, or over-damped

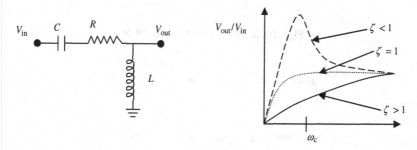

Figure 3.17　Passive second-order high-pass filter

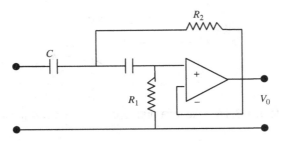

Figure 3.18 Active high-pass high-order filter

when $\zeta > 1$. The order of the filter is determined by the number of capacitors and inductors. An ideal filter requires an nth-order circuit where n is infinity since a higher order leads to a better response, but this increases the cost and complexity of the filter. A trade-off would determine the order and type of filter. Figure 3.18 shows an active high-pass high-order filter.

3.4.3 Band-pass Filter

A band-pass filter passes a range of frequencies while attenuating those both above and below that range. Figure 3.19 illustrates the gain of the filter as a function of the frequency. The upper and lower frequencies determine the band-pass filter. A band-pass filter could be derived from a low-pass filter by substituting C and L in the low-pass filter with the LC circuit as presented in Figure 3.20. The value of L

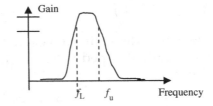

Figure 3.19 Band-pass filter

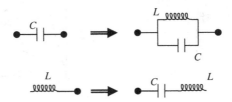

Figure 3.20 Band-pass filter from a low-pass filter transformation

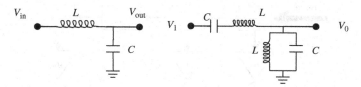

Figure 3.21 Block diagram of a band-pass filter that was transfromed from a low pass filter

and C depends on the pass-band requirements. Figure 3.21 shows an example of a low-pass filter that was transformed to a band-pass filter. The following the values L and C depend on band-pass requirements.

3.5 Digital Filtering

There are many advantages in using digital signal processing (DSP) rather than analog processing. For instance, from a reliability point of view, analog system performance degrades due to aging or temperature, sensitivity to voltage instability, batch-to-batch component variation, high discrete component count and interconnection failures. Digital systems offer no short- or long-term drifts, relative immunity to minor power supply variations, virtually identical components, integrated circuit long life and low development costs. Nevertheless, they are susceptible to single upset events and other radiation issues where a bit flip or corruption can occur. DSPs are widely used in various areas such as radar systems for clutter suppression, matched filters and target tracking and identification. They are also used in image processing for image compression, filtering, enhancement, spectral analysis and pattern recognition. Using discrete logic, digital filters can be built from chain of delay elements. These elements are commonly formed by D-type flip flop. Figure 3.22 shows a typical structure for a digital filter. This structure is constructed using the signal flow shown in Figure 3.23.

The purpose of delaying the signal is to reproduce the frequency response based on the corresponding impulse response. As the number of delay elements increase the response gets closer and closer to the ideal case. This leads to two types of filters: the first one is the finite impulse response (FIR) filter, and the second one is the infinite impulse response (IIR) filter. To see the difference

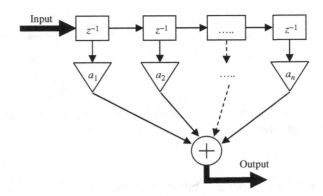

Figure 3.22 Typical digital filtering structure where z^{-1} is the z-transform that represents a delay; a_i is the amplification factor or the filter tap weights

between FIR and IIR, one can look at their impulse responses. The impulse function is defined as

$$\delta_0 = 1; \delta_k = 0; k \neq 0. \tag{3.5.1}$$

Using the FIR system:

$$y_k = x_k + \frac{1}{2}x_{k-1} + \frac{1}{4}x_{k-2}, \tag{3.5.2}$$

the impulse response of this system is

$$y_0 = 1; y_1 = \frac{1}{2}; y_2 = \frac{1}{4}; y_3 = y_3 = \cdots = 0. \tag{3.5.3}$$

For the IIR system:

$$y_k = x_k + ay_{k-1}. \tag{3.5.4}$$

The structures of the two filters are represented in Figure 3.24.

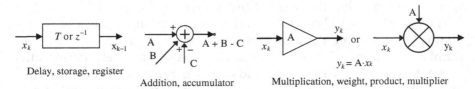

Figure 3.23 Typical signal flow construction

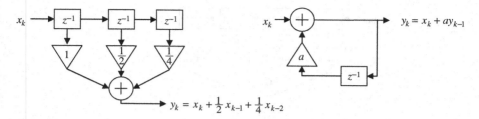

Figure 3.24 Filter structure representation

The impulse response of this system is

$$y_0 = 1; \; y_1 = a; \ldots; \; y_n = a^n. \qquad (3.5.5)$$

This shows that this system has an infinite impulse response IIR and it will only decay if $a < 1$. Figure 3.25 shows the impulse response of the system represented in Figure 3.24.

In general, the transfer function H for an FIR filter is

$$y_k = a_0 x_k + \cdots + a_n x_{k-n} \Rightarrow Y(z)$$
$$= (a_0 + \cdots + a_n z^{-n}) X(z) \Rightarrow H(z) = \sum_{i=0}^{n} a_i z^{-i} \qquad (3.5.6)$$

where the coefficients (a_i) are called filter tap weight. This transfer function consists of zeros only, there are no poles. For recursive or IIR filters

$$y_k = a_0 x_k + \cdots + a_m x_{k-m} - b_1 y_{k-1} - \cdots - b_n y_{k-n};$$
$$H(z) = \frac{a_0 + a_1 z^{-1} + \cdots + a_m z^{-m}}{1 + b_1 z^{-1} + b_2 z^{-2} + \cdots + b_n z^{-n}}. \qquad (3.5.7)$$

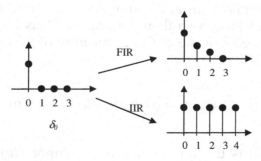

Figure 3.25 FIR versus IIR where $a = 1$

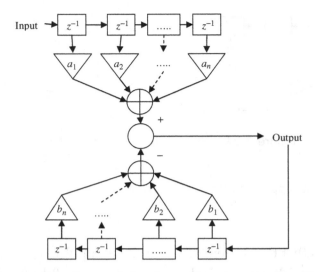

Figure 3.26   General structure of a discrete-time filter

$H(z)$ is rational, with mth-order polynomial as numerator and nth-order polynomial as denominator, and has m zeros and n poles in the z-plane. Factorization of the transfer function $H(z)$: m zeros $z_0, z_1, \ldots,$ z_{m-1}, n poles $p_0, p_1, \ldots, p_{n-1}$

$$H(z) = Cz^{N-M}\frac{(z - z_0)(z - z_1)\cdots(z - z_{m-1})}{(z - p_0)(z - p_1)\cdots(z - p_{n-1})}. \qquad (3.5.8)$$

Figure 3.26 shows the general structure of H. IIR transfer function has both poles and zeros in the z-plane. The frequency response is defined by the location of zeros only. For a given complexity, the number of multiplications and summations, the transition band will be smoother but less sharp than with an IIR. This transfer function could be expressed in terms of the z-transform of the impulse $h(n)$, in that

$$H(z) = \sum h(n)z^{-n} \Rightarrow y(n) = \sum_{r=0}^{\infty} h(r) \cdot x(n - r) = h(n) \otimes x(n). \quad (3.5.9)$$

The output $y(n)$ is the convolution of the input $x(n)$ and impulse response $h(n)$. The difference equation is derived from the knowledge of underlying physics. The impulse response results from the measurement.

For the case where b_i is equal to zero the system is called non-recursive. When b_i is different from zero then the system is called recursive, since the output depends on the past and the future output value, which is the case of IIR. H has to be stable and the polynomial order should be the smallest to minimize complexity whether in parts or in computation and therefore in power consumption. For instance, looking at the example:

$$y_k = x_k + ay_{k-1} \Rightarrow y_n = a^n$$
$$Y(z) = X(z) + az^{-1}Y(z)$$
$$\frac{Y(z)}{X(z)} = \frac{1}{1 - az^{-1}} = \frac{z}{z - a}.$$

(3.5.10)

The magnitude of the pole a determines the stability. If its magnitude is more than one then the impulse response increases indefinitely. If its magnitude is less than one the response decreases exponentially to zero.

The advantages in using an FIR filter are that it can be designed with exact linear phase and that the filter structure is stable with quantized coefficients. Unfortunately, the order of an FIR filter, in most cases, is considerably higher than the order of an equivalent IIR filter meeting similar specifications, which leads to higher computational complexity.

3.5.1 Window Design

There are mainly three methods for the design of FIR filters: the window method, the frequency sampling method and the computer-based method. Here the window method will be explored. For a given filter specification one can choose the filter length m, and a window function $w(n)$, to obtain the transition bandwidth that is the narrowest main-lobe width and the smallest ripple or side-lobe attenuation. There are other types of window functions such as Bartlett (triangular), Blackman, Hanning, Hamming, Kaiser and more (Figure 3.27). Table 3.1 shows some common window functions with their frequency domain characteristics. Hamming windows are by far the most popular.

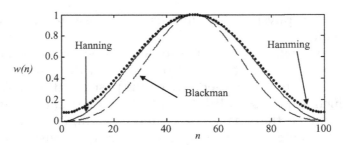

Figure 3.27 Typical window functions[†]

To satisfy filter design requirements one can use an increased number of delays or just apply a specific window. Figure 3.28 shows how to use different windows to mimic an ideal low-pass filter.

Figure 3.29 shows the impact of using a higher-order window. The higher the order the higher the fidelity in mimicking the ideal filter. Unfortunately the complexity increases as well. A trade-off between both is always needed. The attenuation of the filter is often specified,

Table 3.1 Common window functions with some frequency domain characteristics (MLW is the main-lobe width (transition-bandwidth), and the PSB is the peak stop-band attenuation in dB).[25]

Window function	Equation	MLW	PSB
Rectangular	$w[n] = \begin{cases} 1, n \in [0, m-1] \\ 0, \text{otherwise} \end{cases}$	$1.8\pi/m$	-21
Bartlett (triangular)	$w[n] = \begin{cases} 2n/(m-1), n \in [0, (m-1)/2] \\ 2 - 2n/(m-1), n \in [(m+1)/2, m-1] \\ 0, \text{otherwise} \end{cases}$	$6.1\pi/m$	-25
Hanning	$w[n] = \begin{cases} 0.5 - 0.5 \cos(n \cdot 2\pi/(m-1)), n \in [0, m-1] \\ 0, \text{otherwise} \end{cases}$	$6.2\pi/m$	-44
Hamming	$w[n] = \begin{cases} 0.54 - 0.46 \cos(n \cdot 2\pi/(m-1)), n \in [0, m-1] \\ 0, \text{otherwise} \end{cases}$	$6.6\pi/m$	-53
Blackman	$w[n] = \begin{cases} 0.42 - 0.5 \cos(n \cdot 2\pi/(m-1)) + 0.08 \cos(4\pi n/(m-1)), n \in [0, m-1] \\ 0, \text{otherwise} \end{cases}$	$11\pi/m$	-74

[†]Figure 3.27 was generated using the following Matlab code. Figure (3); plot (hamming(100), 'k.'); figure (3);hold on; plot(Blackman (100), 'k–'); figure (3); hold on; plot(hanning(100),'k').

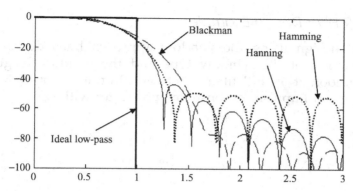

Figure 3.28 Low-pass filter in the frequency domain of various windows of length 20[†]

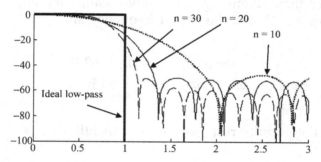

Figure 3.29 Effect of high *n* number, here using a Hamming window[‡]

and therefore the order of the window is identified. A reasonable approximation of the filter order that could be used is

$$n \approx 0.35A/|\omega_s - \omega_p| \qquad (3.5.11)$$

where A is the attenuation of the filter in dB. As for any approximation, it is important to test the system to ensure that it complies with the requirements.

[†]Figure 3.28 was generated using the following Matlab code. $n=20$; for $i = 0 : 2$; $b=$fir1 $(n,0.25,$ (i.*(i-1).*(hamming $(n+1))./2+i.*(2-i).*$hanning(n+1)+(1-i).*(2-i). *(blackman(n+1)))); $Hq=q$filt ('fir',{b}); [h,w,units,href]=freqz(Hq);plot($w,20$*log10(abs(h)),'k:'); xhold on;end"
[‡]Figure 3.29 was generated using the following Matlab code. $n=30$;b =fir1($n, 0.25$,hamming(n+1)); Hq=qfilt('fir',{b}); [h,w,units,href] =freqz(Hq);hold on;plot(w,20* log10(abs(h)), 'k:');end"

3.5.2 FIR Filter Design Example

The FIR design using the window approach boils down to the determination of the window type and the window length. For instance, look at an FIR filter where $\omega_p=0.1\pi$, $\omega_s=0.2\pi$, $R_s=-40\text{dB}$, and a desired impulse response low-pass filter with ω_c:

$$h_d[n] = \sin(\omega_c u)/\pi u$$
$$u = n - (m - 1)/2. \tag{3.5.12}$$

Table 3.1 shows that the PSB of Hamming is -53 dB and for the Hanning window the PSB is -44 dB. Hanning's transition bandwidth is narrower than Hamming's. Therefore using Hanning's window, the window length and the cut-off frequency are

$$m = 6.6\pi/(\omega_p - \omega_s) = 62$$
$$\omega_c = (\omega_p + \omega_s)/2 = 0.15\pi. \tag{3.5.13}$$

Therefore the impulse response $h(n)$ of this FIR filter is

$$h[n] = h_d[n] \cdot w_h[n];$$
$$w[n] = \begin{cases} 0.5 - 0.5 \ \cos(n \cdot 2\pi/(61)), n \in [0, 61]. \\ 0, \text{ otherwise} \end{cases} \tag{3.5.14}$$

Figure 3.30 represents this FIR in the time domain and its spectrum in the frequency domain.

3.5.3 IIR Filter Design

The typical approach to designing an IIR filter is to convert the filter specifications into an equivalent analog filter specification. Then determine the transfer function H_{analog} of the analog filter in the Laplace space and then conduct a mapping from the s-domain to the z-domain in order to maintain the properties of the analog frequency response. Once the transfer function H of the filter is determined, the IIR filter is then identified.

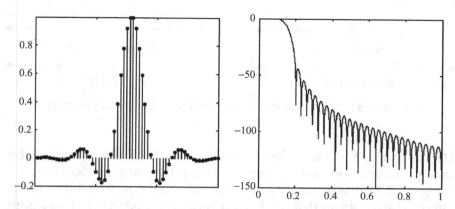

Figure 3.30 FIR in time domain and its spectrum in the frequency domain[†]

3.6 Microwave Filters

Microwave filters operate in the microwave frequency range between 200 MHz and 90 GHz. High frequency signals result in the skin effect which is current on the surface instead of a uniform current distribution in a lumped circuit element. Lumped elements such as resistors, capacitors and inductors tend to behave like antenna elements since their size is comparable to the microwave wavelength. Therefore their use as a circuit element in a filter application becomes undesirable. Noise is generated from the interconnect between the lumped elements. Alternatively, transmission lines for RF rely typically on coaxial cables, microstrips or waveguides. In coaxial cables the electrical field is contained within both conductors, as well as for microstrips. One can imagine that a microstrip is a coaxial cable with an infinite radius. Nevertheless, and because of the heritage in circuit theory, analogy to lumped elements, i.e. inductors and capacitors, and their equivalent circuit could be used to design RF filters.

A microwave filter implementation could be accomplished in a few steps. The first step is to transform the design requirement into an equivalent lumped circuit, then using Richard's transformation, convert lumped elements to transmission line sections. The next

[†]Figure 3.30 was generated using the following Matlab code. $wp=1*pi; ws=.2*pi; m=6.2*pi/(ws-wp); n=0:m-1; wc=(wp+ws)/2; u=n-(m-1)/2; hd=10*sin(wc*u)./(pi*u); wh=$hanning(m)'; $h=hd.*wh$; subplot(121); stem(h/max(h)); $N=1000; [H,w]=$freqz(h,1,N);subplot(122); plot(w/pi,20*log10 (abs(H)/max(abs(H))));

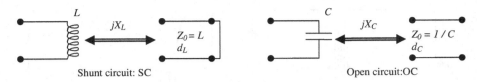

Figure 3.31 Inductance and capacitor circuit equivalent

step is to use Kuroda's identities which separate filter elements by using transmission line sections and simplify complex elements.

Figure 3.31 illustrates how the inductor can be replaced by a short circuited stub of length d_L and characteristic impedance L. This figure also shows how the capacitor can be replaced by an open circuited stub of length d_C and characteristic impedance $1/C$:

$$d_C = CZ_{min}v/R_0f$$
$$d_L = LR_0v/Z_{max}f$$
(3.6.1)

where Z_{max} is the maximum impedance, R_0 is the filter impedance, L is the normalized inductance, C is the normalized capacitance, v is the propagation of light in that media and f_c is the cut-off frequency. For instance, for a low-pass filter with $f_c = 1$ GHz, $Z_{max} = 120$ ohms, $Z_{min} = 12$, $R_0 = 50\ \Omega$, $C = 1$, $L = 2$ then $d_C = 28.5$ mm, $d_L = 8.25$ mm. A lossless microstrip line has L and C per unit length. A short section of microstrip line with a large impedance acts like a series inductor. A short section of microstrip line with a small impedance acts like a shunt capacitor with which one can design a step impedance low-pass filter (Figure 3.32).

Using Kuroda's identity, where $\lambda_c/8$ are redundant lines called commensurate lines to separate stubs, the equivalence between open stubs and shunt stubs is shown in Figure 3.33:

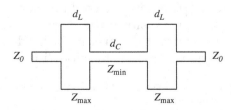

Figure 3.32 Low-pass filter

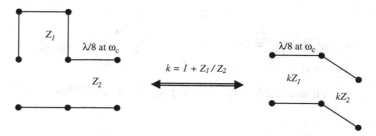

Figure 3.33 Equivalent series stub and shunt stub[19]

Figure 3.34 shows a low level approach that one can use to design a microwave filter. The sequence is to identify the specification of the filter, its lumped circuit equivalent, use Richard's transform, then add redundant $\lambda_c/8$ lines, and then use Kurod's identity to obtain the open stub configuration.[19] The microwave filter's physical characteristics are then identified. Figure 3.35 shows the equivalent of

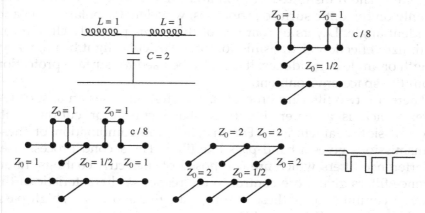

Figure 3.34 Example of lumped/microstrip derivation for a low-pass filters

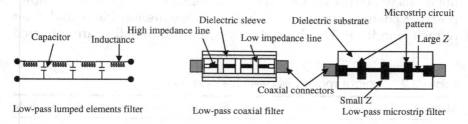

Figure 3.35 Equivalent low-pass filter circuit using lumped elements, coaxial and microstrip[19]

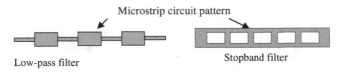

Figure 3.36   Low-pass and stopband filter

microstrip and coaxial filters of lumped elements. Figure 3.36 illustrates some configurations of a low-pass filter and a stopband filter.

3.7. Optical Filters

When a light beam reaches an optic surface such as a mirror, a lens or a beam splitter, part of the light is transmitted and part of it is reflected. The transmitted portion and the reflected portion depend mainly on the optic surface properties, wavelength, polarization and incident angle. Layers of coatings of thin films properly chosen can optimize reflection or transmission of a particular light beam, wavelength or angle of incidence. It can also be used for surface protection from the space environment.

There are two filtering processes. The first one uses an absorption filter which is a material with a sharp cut-off or cut-on in its transmission characteristic. For instance, the combination of these transmissions gives a band-pass like filter. The second process uses interference filters which are composed of dielectric coatings. Interference filters give more control to the response characteristic of the filter. A combination of these coatings can give a low-pass, high-pass, band-pass or notch type of filter.

An optical coating consists of a layer, or series of layers, of different materials which are deposited over the optic surface. Table 3.2 shows possible coating materials with their transmission band and their refractive index n. The desired properties of the coating are achieved

Table 3.2   Typical materials used for coating

Material	Germanium	Sapphire	Zinc selenide	Zinc sulphide
Transmission band (µm)	2–20	0.4–5	0.5–20	0.6–15
Refractive index	4	1.63	2.4	2.2

by a mixture of interference and intrinsic properties of the materials that are used. The thickness of the layer determines the behavior of the coating. Quarter-wave layers give the maximum interference effect. However, half layers have no effect. The reflectance depends on the number of the layer and on the wavelength. For instance, the reflectance of metallic layers increases as the wavelength increases whereas the reflectance of dielectric layers decreases.

For an incident beam on an optic surface of refractive index n_s covered with a coating of refractive index n_l and a thickness h, the normal incidence reflectance R, which is the ratio of the reflected power to the incident power for three media, is[11]

$$R = \frac{n_l^2(n_0 - n_s)^2\cos^2(kh) + (n_0 n_s - n_l^2)^2\sin^2(kh)}{n_l^2(n_0 + n_s)^2\cos^2(kh) + (n_0 n_s + n_l^2)^2\sin^2(kh)}. \qquad (3.7.1)$$

For a quarter wavelength thick coating, the reflectance is

$$kh = \frac{\pi}{2} \Rightarrow h = \frac{\lambda}{4} \Rightarrow R = \frac{(n_0 n_s - n_l^2)^2}{(n_0 n_s + n_l^2)^2}. \qquad (3.7.2)$$

The reflectance is null for

$$n_0 n_s = n_l^2 \Rightarrow R = 0. \qquad (3.7.3)$$

For a half wavelength thick coating, the reflectance is

$$kh = \pi \Rightarrow h = \frac{\lambda}{2} \Rightarrow R = \frac{(n_0 - n_s)^2}{(n_0 + n_s)^2}. \qquad (3.7.4)$$

For multilayer coatings of quarter-wave thickness (as shown in Figure 3.37), with x layers of coating 1 and $x - 1$ layers of coating 2, the reflectance for a normal incidence is

$$R = (n_0 n_s - n_t)/(n_0 n_s + n_T); n_t = n_{l2}^{2x}/n_{l1}^{2x-2} \qquad (3.7.5)$$

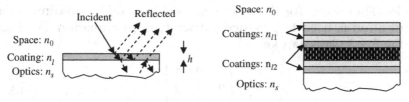

Figure 3.37 One layer and multilayer coating on an optic surface

where n_l, n_t, and n_s, are respectively the refractive indices of the quarter wavelength coating layer, transformed surface and optic surface. The phase shift δ of a traveling wave through thickness d of a thin film is

$$\delta = -2\pi n/\lambda. \qquad (3.7.6)$$

The minus sign indicates a phase delay.

3.8 Digital Image Filtering

Spacebased imaging is susceptible to various type of noise such as white noise or 'salt and pepper noise' known also as impulsive noise. Noise could result from faulty sensor elements, upset, AD conversion noise, optic contamination, aberration, blurring or distortion. Image filtering modifies pixel information in an image using a function that depends on the adjacent pixels. It can also be used to enhance some features or to isolate a specific signal from the surrounding environment for further processing such as pattern recognition:

$$f(i,j) = K \otimes I = \sum_k \sum_l K(i-k, j-l)I(k,l) \qquad (3.8.1)$$

where I is the image and K is the kernel that is used to suppress the noise. The product sign is the convolution sign. Based on the kernel one can either have a low-pass filter, band-pass filter or edge detection among others. For instance, a Gaussian low-pass filter could be used:

$$G = K(i,j) = \frac{1}{2\pi\sigma^2} \exp\left(-\frac{\left((i-k-1)^2 + (j-k-1)^2\right)}{2\sigma^2}\right) \qquad (3.8.2)$$

where K is a $2k+1$ by $2k+1$ array. Images are composed of discrete pixels. Therefore, for Gaussian smoothing, this two-dimensional distribution is used to produce a discrete approximation. For example the 3×3 Gaussian kernel is

$$\frac{1}{16}\begin{bmatrix} 1 & 2 & 1 \\ 2 & 4 & 2 \\ 1 & 2 & 1 \end{bmatrix}. \qquad (3.8.3)$$

The ratio $\frac{1}{16}$ is the normalization factor. For the 5×5 an approximation to the normalized Gaussian kernel is

$$\frac{1}{273}\begin{bmatrix} 1 & 4 & 7 & 4 & 1 \\ 4 & 16 & 26 & 16 & 4 \\ 7 & 26 & 41 & 26 & 7 \\ 4 & 16 & 26 & 16 & 4 \\ 1 & 4 & 7 & 4 & 1 \end{bmatrix}. \qquad (3.8.4)$$

If object reflectance changes slowly and noise at each pixel is independent, then one can replace each pixel with the average of the neighbors. This noise filtering is called mean smoothing or box filter, which corresponds to the *sinc* function in the frequency domain. This would therefore act as a low-pass filter, and

$$K = \frac{1}{9}\begin{bmatrix} 1 & 1 & 1 \\ 1 & 1 & 1 \\ 1 & 1 & 1 \end{bmatrix}. \qquad (3.8.5)$$

3.9 Kalman Filter

The least-squares method is a very common estimation method. The parameter estimates are the values that minimize the sum of the squares of the differences between the observations and modeled values based on parameter estimates. The Kalman filter is an asset of mathematical equations that provides an efficient recursive computational solution of the least-squares method using the state model process. The filter is very powerful in several aspects: it supports estimations of past, present and even future states, and it can do so even when the precise nature of the modeled system is unknown. Kalman filters are widely used in GPS receivers to estimate velocity and position as a function of time, in clocking because of their white noise-like behaviors, in atmospheric delay variations, satellite orbits and many more. In the following section an introduction to state variables, followed by the discrete Kalman filter, will be covered.

3.9.1 State-space Representation

The basic *RLC* circuit is shown in Figure 3.38. The following differential equations are derived using Kirchoff's law

$$I_C = C\partial_t V_C = I_s - I_L$$
$$L\partial_t I_L = -RI_L + V_C \qquad (3.9.1)$$
$$V = RI_L$$

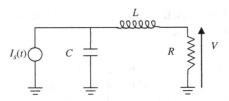

Figure 3.38   *RLC circuit*

where I_C, I_L, and I_s are the capacitor current, the inductor current and
the source current respectively. V_C and V are the capacitor voltage,
and the output voltage. Using a set of variables judicially chosen x_1
and x_2, called state variables, the previous set of equations becomes

$$\begin{cases} x_1 = V \\ x_2 = I_L \\ y_1 = V_C \end{cases} \Rightarrow \begin{cases} \partial_t x_1 = I_s/C - x_2/C \\ \partial_t x_2 = x_1/L - Rx_2/L. \\ y_1 = Rx_2 \end{cases} \qquad (3.9.2)$$

Using the following variables:

$$X = \begin{pmatrix} x_1 \\ x_2 \end{pmatrix}; Y = \begin{pmatrix} y_1 \\ y_2 \end{pmatrix}; A = \begin{pmatrix} 0 & -1/C \\ 1/L & -R/L \end{pmatrix}; B = \begin{pmatrix} 1/C \\ 0 \end{pmatrix}.$$

$$u = I_s; C = (0 \ R); D = 0 \qquad (3.9.3)$$

This system could be rewritten:

$$\begin{cases} \dot{X} = AX + Bu \\ Y = CX + Du \end{cases}. \qquad (3.9.4)$$

The combination of the state equation and the output equation is
called the state model.

3.9.2 Discrete Kalman Filter

For the discrete Kalman filter, the state model is conventionally
written as

$$\begin{cases} X_{k+1} = \phi_k X_k + W_k \\ Z_k = H_k X_k + V_k \end{cases} \qquad (3.9.5)$$

where X_k is the process vector at time t_k, ϕ_k is the state transition at
time t_k, W_k is the process noise, Z_k is the measurement vector at time

t_k, H_k is the noiseless transfer function between the measurement and the state vector and V_k is the measurement error. Parameters with no process noise are called deterministic. For this introduction, V_k and W_k are uncorrelated with a known covariance structure:

$$E(W_i W_k^T) = \delta_{ik} Q_k$$
$$E(V_i V_k^T) = \delta_{ik} R_k$$
$$E(W_i V_k^T) = 0 \qquad (3.9.6)$$
$$\delta_{ik} \begin{cases} i = k \Rightarrow \delta_{ii} = 1 \\ i \neq k \Rightarrow \delta_{ik} = 0 \end{cases}$$

where $W_k{}^T$ is the transpose of W_k. Assuming that

$$\left. \begin{array}{l} \xi_k^- = X_k - \hat{X}_k^- \\ P_k^- = E(\xi_k^- \xi_k^{-T}) \\ P_k = E(\xi_k \xi_k^T) \end{array} \right\} \Rightarrow P_k = (I - K_k H_k) P_k^- \qquad (3.9.7)$$

where $^\wedge$ refers to the estimate, the minus superscript refers to the best a priori estimate of the measurement conducted at time t_k, ξ_k^- is the estimate error with a zero mean, and P_k^- is the best estimate of the covariance error matrix prior to the measurement t_k and P_k is the error covariance matrix at t_k. The updated estimate is then:

$$\hat{X}_k = \hat{X}_k^- + K_k \left(Z_k - H_k \hat{X}_k^- \right) \qquad (3.9.8)$$

where K_k is the Kalman gain. The Kalman gain is therefore:

$$K_k = P_k^- H_k^T \left(H_k P_k^- H_k^T + R_k \right)^{-1}. \qquad (3.9.9)$$

The $^{-1}$ refers to the inverse of the sum of the matrices. Since W_k has a zero mean and it is uncorrelated with a priori W, then using equation (3.9.5):

$$\hat{X}_{k+1}^- = \phi_k \hat{X}_k \Rightarrow \xi_k^- = \phi_k \xi_k^- + W_k \Rightarrow P_{k+1}^- = \phi_k P_k \phi_k^T + Q_k. \qquad (3.9.10)$$

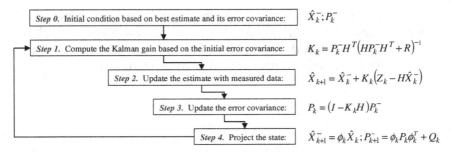

Figure 3.39 Kalman filter steps – the basic algorithm of the discrete Kalman filter[3]

These equations constitute the Kalman filter recursive equations. The Kalman filter process is then summarized in Figure 3.39.

For example, looking at the following system

$$x_k = x_{k+1}$$
$$z_k = x_k + v_k \tag{3.9.11}$$

where x describes the system and z describes the measurement. This system shows that x is constant. However, the measurement could be corrupted by a noise. Assuming that this noise is uncorrelated with zero mean and variance σ^2

$$E(v_k) = 0$$
$$E(v_k^2) = \sigma^2. \tag{3.9.12}$$

Therefore, using Figure 3.39, the Kalman filter sequence becomes Figure 3.40.

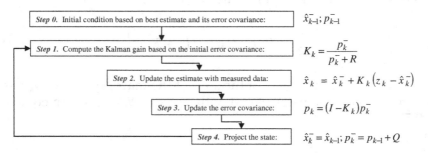

Figure 3.40 Kalman filter steps for this scalar model

Using these steps the error covariance takes the following values

$$p_1 = \frac{\sigma^2 p_0}{\sigma^2 + p_0}$$

$$p_2 = \frac{\sigma^2 p_1}{\sigma^2 + p_1} = \frac{\sigma^2 p_0}{\sigma^2 + 2p_0} \qquad (3.9.13)$$

$$p_n = \frac{\sigma^2 p_0}{\sigma^2 + np_0}.$$

Therefore the Kalman gain is

$$K_k = \frac{p_0}{\sigma^2 + kp_0}. \qquad (3.9.14)$$

The updated estimate with the measured data becomes

$$\hat{x}_k = \hat{x}_k^- + \frac{p_0}{\sigma^2 + kp_0}\left(z_k - \hat{x}_k^-\right). \qquad (3.9.15)$$

For large k, this equation becomes

$$\hat{x}_k \approx \hat{x}_k^-. \qquad (3.9.16)$$

This would indicate that more measurement will not improve the value of x.

Bibliography

1. Bendat, J. S. and Piersol, A. G. (1971) *Random Data: Analysis and Measurement Procedures*, Wiley-Interscience.
2. Blake, R. (2002) *Electronic Communication Systems*, Delmar.
3. Bozic, S. M. (1979) *Digital and Kalman Filtering*, John Wiley and Sons.
4. Brown, R. G. (1983) *Introduction to Random Signal Analysis and Kalman Filtering*, John Wiley and Sons.
5. Dereniak, E. L. and Boreman, G. D. (1996) *Infrared Detectors and Systems*, John Wiley and Sons, Inc.
6. Dorf, R. C. and Bishop, R. H. (2001) *Modern Control Systems*, 9th edn, Prentice Hall.
7. Forsyth, D. A. and Ponce, J. (2003) *Computer Vision: A Modern Approach*, Prentice-Hall.
8. Fraden, J. (1997) *Handbook of Modern Sensors*, American Institute of Physics.

9. Freeman, B. and Abid, M. (2003) *Out of Range Cost Estimating using a Kalman Filter*, ISPA/SCEA.
10. Gonzalez, R. and Woods, R. (1992) *Digital Image Processing*, Addison-Wesley Longman Publishing Co., Inc.
11. Hecht, E. (1998) *Optics*, 3rd edn, Addison-Wesley.
12. Holman, J. P. (1994) *Experimental Methods for Engineers*, McGraw-Hill, Inc.
13. Jones, N. B. and Watson, J. D. M. (1990) *Digital Signal Processing: Principles, Designs and Applications*, Peter Peregrinus Ltd.
14. Jelalian, A. V. (1992) *Laser Radar Systems*, Artech House.
15. Kingston, R. H. (1995) *Optical Sources Detectors and Systems, Fundamentals and Applications*, Academic Press Inc.
16. Mohanty, N. (1991) *Space Communication and Nuclear Scintillation*, Van Nostrand Reinhold.
17. Osche, G. R. (2002) *Optical Detection Theory for Laser Application*, John Wiley and Sons Inc.
18. Pease, C. B. (1991) *Satellite Imaging Instruments*, Ellis Horwood Limited.
19. Pozar, D. M. (1990) *Microwave Engineering*, Addison Wesley.
20. Sloggett, D. R. (1989) *Satellite Data Processing, Archiving and Dissemination*, Ellis Horwood Limited.
21. Taylor, H. (1997) *Data Acquisition for Sensor Systems*, Chapman and Hall.
22. Trucco, E. and Verri, A. (1998) *Introductory Techniques for 3-D Computer Vision*, Prentice Hall.
23. Williams, A. B. and Taylor, F. J. (1995) *Electronic Filter Design Handbook*, McGraw-Hill Inc.
24. Winder, S. (2002) *Analog and Digital Filter Design*, Newnes.
25. Young, T. (1985) *Linear Systems and Digital Signal Processing*, Prentice-Hall Inc.

4

Infrared Sensors

An optical sensor may collect infrared (IR) light coming from a source of interest. This light undergoes a process that is summarized in Figure 4.1. First, the light is collected using static or dynamic optics such as scanning mechanisms and collimated to fit the subsequent optics. The front optics usually use an antireflection coating to ensure that all light of interest goes into the sensor. Further filters and collimating optics are used to select subsets of wavelengths known as spectral bands which pass through and focus onto a detector (or a set of detectors) that transforms photons to voltage. This voltage is amplified, digitized and then processed. This transformation has some properties that are unique to IR sensors, such as cooling the detector to minimize noise and using exotic materials to enhance the photon electron conversion. Included in this chapter is a review of governing equations of electromagnetic fields, optics and detectors. The reviewed subjects will be used to understand some spacebased IR examples.

4.1 Electromagnetic Waves

4.1.1 The Electromagnetic Spectrum

The electromagnetic spectrum ranges from the shorter wavelengths, gamma and X-rays, to the longer wavelengths, microwaves and broadcast radio waves. The spectrum is presented in Figure 4.2. The visible portion of the spectrum represents only a small part of

Spacecraft Sensors.   Mohamed M. Abid
© 2005 John Wiley & Sons, Ltd

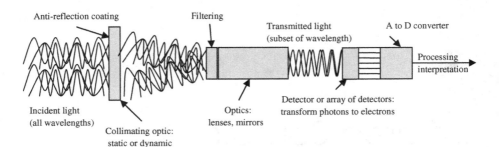

Figure 4.1 Basic steps of optical/IR detectors

the spectrum. In order to detect the rest of the spectrum, sensors have to be used. The visible wavelengths cover a range from approximately 0.4 μm (violet) to 0.7 μm (red). When sunlight passes through a prism, its initial apparently uniform or homogenous color separates into a spectrum of colors.

The IR region covers the wavelength range from approximately 0.7 μm to 1000 μm. It can be divided into two categories based on radiation properties as illustrated in Figure 4.3. The first category, from 0.7 μm to 3 μm, is the reflected IR region which is used for remote sensing purposes in ways very similar to radiation in the visible portion. The second category, from 3 μm to 1000 μm, is the

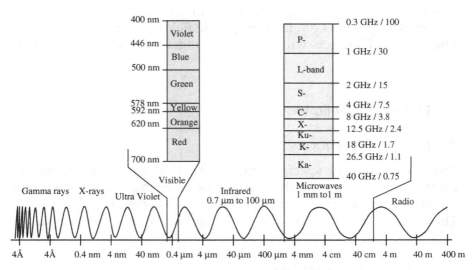

Figure 4.2 The electromagnetic spectrum

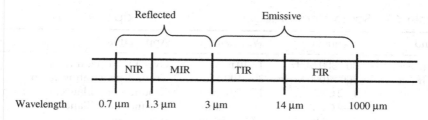

Figure 4.3 The infrared Spectrum

emitted IR region which is quite different from the reflected IR portions as this energy is essentially emitted radiation in the form of heat.

Within the reflected IR category two subcategories are distinguished based on the sensing element. The first subcategory is the near IR (NIR), from 0.7 μm to 1.3 μm, which can be recorded photographically using special false color photographic films and charged coupled devices (CCDs). NIR imagery is used in soil analysis, agricultural studies and vegetation studies. The second subcategory is the mid IR (MIR), ranging from 1.3 μm to 3 μm, and this can be detected with a specially designed thermal sensor or a radiometer.

Emissive IR extends from 3 μm to 1000 μm. Within the emissive IR there are two subcategories. There is the thermal IR (TIR), which ranges from 3 μm to 14 μm, and the far IR (FIR), which ranges from 14 μm to 1000 μm. The simplest systems that use TIR are video based and have been used successfully in low cost remote sensing systems. Table 4.1 presents typical spacebased applications that are associated with each band.

4.1.2 Maxwell's Equations

The following sets of equations are called Maxwell's equations. They describe the electromagnetic waves' behavior in various media. An ensemble of electric charges generates electric fields and the field lines diverge from, or converge toward, the charges. This is summarized using Gauss' Law of electricity:

$$\nabla \bullet D = \partial_x D_x + \partial_y D_y + \partial_z D_z = \rho$$
$$\partial_x D_x \equiv \partial D_x / \partial X$$

(4.1.1)

where D is the electric displacement or electric flux density vector, ρ is the electric charge density and ∇ is the divergence operator which

Table 4.1 Some electromagnetic earth based applications

Band	Frequency	Wavelength	Application
VHF	30–300 MHz	1–10 m	Over the horizon radar
UHF	300–1000 MHz	30–100 cm	Ground penetrating radar
L	1–2 GHz	15–30 cm	Ground surveillance, astronomy
S	2–4 GHz	75–150 mm	Ground surveillance
C	4–8 GHz	37.5–75 mm	Spacebased SAR
X	8–12.5 GHz	24–37.5 mm	Fire control radar, proximity
Ku	12.5–18 GHz	16.7–24 mm	Collision avoidance
K	18–26.5 GHz	11.3–16.7 mm	Fire control radar
Ka	26.5–40 GHz	7.5–11.3 mm	Surveillance
Millimetre	30–300 GHz	1–10 mm	Astronomy
Sub–mm		50–1 mm	Astronomy
Far IR		14–50 mm	Molecular properties
Longwave IR		8–14 mm	Laser radar
Near IR		1–3 mm	Personnel detection
Very near IR		760–1000 nm	Imaging, laser ranging
Visible		380–760 nm	Imaging
Ultraviolet		100–380 nm	Missile plume detection

is the spatial vector derivative operator. There are no sources of magnetic fields that are analogous to electric charges, this is summarized in Gauss' law of magnetism:

$$\nabla \bullet B = \partial_x B_x + \partial_y B_y + \partial_z B_z = 0 \qquad (4.1.2)$$

where B is the magnetic flux density. Electric fields can be generated by time-varying magnetic fields and electric fields generated in this way tend to coil or curl around the magnetic field lines. This is expressed with Faraday's Law of induction:

$$\nabla \times E = \begin{pmatrix} \partial_y E_z - \partial_z E_y \\ \partial_z E_y - \partial_y E_z \\ \partial_x E_y - \partial_y E_x \end{pmatrix} = -\frac{\partial B}{\partial t} \qquad (4.1.3)$$

where E is the electric field intensity, t is time, and $\nabla \times$ is the curl operator or the spatial vector derivative operator. A change in the electrical field causes a change in the magnetic field. Magnetic fields can be generated by time-varying electric fields and the magnetic fields tend to curl around these sources. This is expressed using Ampère's Law:

$$\nabla \times H = \begin{pmatrix} \partial_y H_z - \partial_z H_y \\ \partial_z H_y - \partial_y H_z \\ \partial_x H_y - \partial_y H_x \end{pmatrix} = J + \partial_t D \qquad (4.1.4)$$

where, H is the magnetic field intensity and J is the electric current density.

The magnetic fields generated by currents and calculated from Ampère's Law are characterized by the magnetic field B. However, when the generated fields pass through magnetic materials, which themselves contribute to internal magnetic fields, ambiguities can arise about what part of the field comes from the external currents and what comes from the material itself. Therefore the magnetization of the material M is introduced:

$$B = \mu_0(H + M). \qquad (4.1.5)$$

When there is no material magnetization, the relationship between B and H is

$$B = \mu H$$
$$\mu = \mu_r \mu_0 \qquad (4.1.6)$$

μ_0 being the magnetic permeability of space and μ_r the relative permeability of the material. If the material does not respond to the external magnetic field by producing any magnetization, then $\mu_r = 1$. For real media the index of refraction is

$$v = \frac{\omega}{k} = \frac{1}{\sqrt{\varepsilon\mu}}$$

$$c = \frac{1}{\sqrt{\varepsilon_0 \mu_0}}$$

$$\varepsilon_r = \frac{\varepsilon}{\varepsilon_0} \qquad (4.1.7)$$

$$\mu_r = \frac{\mu}{\mu_0}$$

$$n = \frac{c}{v} = \sqrt{\varepsilon_r \mu_r},$$

where v is the speed of light in real media, c is the speed of light in free space, μ is the magnetic permeability of the medium which is the ability to store or conduct a magnetic field, ϵ is the electric permittivity of the medium which is the ability to store or conduct an electric field, ϵ_r is the relative permittivity (dielectric constant), μ_r is the relative permeability and n is the index of refraction which is the ratio of the speed of light in a medium relative to the speed of light in a vacuum. A more general form of the index of refraction expressed as a complex index of refraction is

$$m = n + i\alpha\lambda/4\pi \tag{4.1.8}$$

where n is the real index of refraction and α is the absorption coefficient. The dielectric constant ϵ_r is

$$\epsilon_r = \epsilon' + i\sigma/\omega\epsilon_0 \tag{4.1.9}$$

where σ is the conductivity. These equations are very valuable, for example, in ground penetrating radars. Finally, the complex permittivity and the complex index of refraction are related by

$$\epsilon = m^2. \tag{4.1.10}$$

In free space, Maxwell's equations become

$$
\begin{aligned}
\rho &= 0 \\
j &= 0 \\
\nabla \cdot E &= 0 \\
\nabla \cdot B &= 0 \\
\nabla \times E &= -\partial_t B \\
\nabla \times B &= \epsilon_0\mu_0\partial_t E
\end{aligned}
\tag{4.1.11}
$$

$$
\begin{aligned}
D &= \epsilon_0 E \\
B &= \mu_0 H.
\end{aligned}
\tag{4.1.12}
$$

Using the previous equations, the wave equation of the electric field is

$$
\left.
\begin{aligned}
\nabla \times (\nabla \times E) &= -\nabla \times (\partial_t B) = -\partial_t(\epsilon_0\mu_0\partial_t E) = -\epsilon_0\mu_0\partial_t^2 E \\
\nabla \times (\nabla \times E) &= \nabla(\nabla \cdot E) - \nabla^2 E = -\epsilon_0\mu_0\partial_t^2 E
\end{aligned}
\right\}
\Rightarrow \nabla^2 E = \frac{1}{c^2}\partial_t^2 E.
\tag{4.1.13}
$$

4.1.3 Wave Equation

The one-dimensional wave equation has the following form:

$$\frac{\partial^2 \psi}{\partial x^2} = \frac{1}{v^2}\frac{\partial^2 \psi}{\partial t^2}$$

(4.1.14)

where ψ is an arbitrary function and v is the velocity. Light waves or their electric field will be a solution to this equation. If $f_1(x,t)$ and $f_2(x,t)$ are solutions to the wave equation, then $f_1(x,t) + f_2(x,t)$ is also a solution. This is known as the principle of superposition. It also means that waves can constructively or destructively interfere. The one-dimensional wave equation has the simple solution of

$$\psi(x,t) = f(x \pm vt)$$

(4.1.15)

where v is the wave velocity. This can be shown using the change of variable $u = x - vt$ and the wave equation. This solution indicates that the wave shape is conserved for either a traveling wave in the $+v$ x-direction and in the $-v$ x-direction. A general solution of the above equation is

$$\psi(x,t) = \Re\left(Ae^{i\omega(t-x/v)}\right) = \Re\left(Ae^{i(\omega t-kx)}\right)$$

(4.1.16)

where k is the wave-number, $k = 2\pi/\lambda$, $v = \omega/k$ is the phase velocity of the wave and $\Re$ is the real part of the complex function. For an ideal transmission line, one can derive the wave equations that describe the potential or the current in the line. A transmission line could be represented by a lumped element circuit (Figure 4.4) i.e. a circuit composed of capacitors, inductors and resistors.

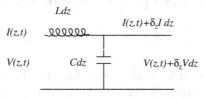

Figure 4.4 The equivalent circuit for a transmission line

Therefore, the governing equation of the voltage and the current are

$$\frac{\partial^2 V(z,t)}{\partial z^2} - \frac{1}{v^2}\frac{\partial^2 V(z,t)}{\partial t^2} = 0$$

$$\frac{\partial^2 I(z,t)}{\partial z^2} - \frac{1}{v^2}\frac{\partial^2 I(z,t)}{\partial t^2} = 0 \qquad (4.1.17)$$

$$v = 1/\sqrt{LC}.$$

For a wave propagating in the $+v$ x-direction:

$$\psi(x,y,z,t) = A(y,z)e^{j(\omega t - kx)}. \qquad (4.1.18)$$

If $A(y,z)$ is constant, the resulting wave is called a plane wave. Surfaces of constant phase, $\omega t - kx = 0$, are called wavefronts. They are locally perpendicular to the direction of propagation.

The general three-dimensional wave equation is

$$\nabla^2 \psi = \frac{\partial^2 \psi}{\partial x^2} + \frac{\partial^2 \psi}{\partial y^2} + \frac{\partial^2 \psi}{\partial z^2} = \frac{1}{v^2}\frac{\partial^2 \psi}{\partial t^2}. \qquad (4.1.19)$$

The general plane wave solution is

$$\psi(r,t) = Ae^{i(\omega t - k.r)}. \qquad (4.1.20)$$

For propagation in an arbitrary direction defined by a unit vector **n**, the wave vector is defined as

$$k = k \cdot n = k_x x + k_y y + k_z z$$

$$k^2 = k_x^2 + k_y^2 + k_z^2 = (2\pi/\lambda)^2. \qquad (4.1.21)$$

For spherical waves, the waves expanding symmetrically from a point source are described by the following equation:

$$\psi(r,t) = \frac{A}{r}e^{i(\omega t - kr)}. \qquad (4.1.22)$$

Conservation of energy leads to a decay of the amplitude with distance from the source. Since energy is proportional to $|\psi|^2$, energy flowing across a sphere of radius r is proportional to $4\pi r^2 |\psi|^2$ and

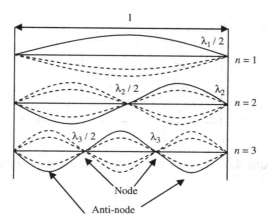

Figure 4.5 Confined one-dimensional wave between two boundaries

therefore $\psi\mu_1/r$. For large r, this spherical wave approximates a plane wave.

Standing waves arise whenever there are two identical counter-propagating waves. Examples would include a stretched string, a laser cavity and quantum wells. The resulting wave becomes

$$\psi = Ae^{i(\omega t - kx)} - Ae^{i(\omega t + kx)} = 2A\sin(kx)\sin(\omega t). \qquad (4.1.23)$$

When the wave is confined between two boundaries, as illustrated in Figure 4.5, and the wave at the boundary is null, the quantization condition is

$$x = l, \psi = 0 \Rightarrow \psi = 2A\sin(kl)\sin(\omega t) = 0 \Rightarrow k = n\pi/l. \qquad (4.1.24)$$

The two-dimensional or three-dimensional solution of the wave equation can be used in many systems such as electron states, potential wells, black-body radiation, waveguides and others.

Looking at the two-dimensional standing wave as illustrated in Figure 4.6, in a bounded volume (a, b, ∞) where each boundary results in a reflection, the wave equation solution is

$$
\begin{aligned}
\psi_1 &= A\exp(i\omega t - ik_x x - ik_y y)\\
\psi_2 &= -A\exp(i\omega t - ik_x x + ik_y y)\\
\psi_3 &= A\exp(i\omega t - ik_x x + ik_y y)\\
\psi_4 &= -A\exp(i\omega t - ik_x x - ik_y y).
\end{aligned}
\qquad (4.1.25)
$$

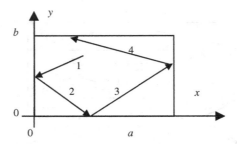

Figure 4.6 Two-dimensional bounded volume

Each reflection results in a phase π change in the wave. Quantization conditions are obtained using the superposition principle and boundary conditions:

$$\psi = \psi_1 + \psi_2 + \psi_3 + \psi_4 = -4Ae^{ii\omega t}\sin k_y y \sin k_x x$$
$$k_x = n_x\pi/a \tag{4.1.26}$$
$$k_y = n_y\pi/b.$$

Each pair of integers (n_x, n_y), specifies a normal mode. The frequency of vibration follows from

$$\left.\begin{array}{l} k^2 = k_x^2 + k_y^2 \\ \omega = vk \end{array}\right\} \Rightarrow \omega = v\pi\sqrt{n_x^2/a^2 + n_y^2/b^2} \tag{4.1.27}$$

The quantization conditions in three-dimensions give us

$$k_x = \frac{n_x\pi}{a}$$
$$k_y = \frac{n_y\pi}{b} \tag{4.1.28}$$
$$k_z = \frac{n_z\pi}{c}$$

$$\left.\begin{array}{l} k^2 = k_x^2 + k_y^2 + k_z^2 \\ \omega = vk \end{array}\right\} \Rightarrow \omega = v\pi\sqrt{n_x^2/a^2 + n_y^2/b^2 + n_z^2/c^2} \tag{4.1.29}$$

where ω is the frequency of the normal modes.

4.1.4 Solution to Maxwell's Equations

Applying the previous analysis and results to Maxwell's equation for a plane wave gives the solution:

$$
\begin{aligned}
E_x &= E_0 \cos(\omega t - kz) \\
E_y &= E_z = 0 \\
B_y &= E_0 \cos(\omega t - kz)/c \\
B_x &= B_z = 0 \\
c &= \omega/k = 1/\sqrt{\varepsilon_0 \mu_0}
\end{aligned}
\tag{4.1.30}
$$

where c is the speed of light in free space, f is the frequency in Hz, λ is the wavelength and k is the wave number. The electromagnetic wave is an oscillating electric field E and magnetic field B. They are transverse waves propagating at a velocity v. In a vacuum $v = c \sim 3 \times 10^8$ m/s.

$$
v = E \times B.
\tag{4.1.31}
$$

The three-dimensional Maxwell's equation for the electric field is

$$
\nabla^2 E = \varepsilon \mu \partial_t^2 E \Rightarrow \frac{\partial^2 E}{\partial x^2} + \frac{\partial^2 E}{\partial y^2} + \frac{\partial^2 E}{\partial z^2} = \varepsilon \mu \frac{\partial^2 E}{\partial t^2} = \frac{1}{v^2} \frac{\partial^2 E}{\partial t^2}.
\tag{4.1.32}
$$

A solution of these equations is in the form:

$$
\begin{aligned}
E &= E_0 \exp(i(k \cdot r - \omega t)) \\
k \cdot r &= k_x x + k_y y + k_z z.
\end{aligned}
\tag{4.1.33}
$$

The energy density of an electric field U_E, and magnetic field U_B are

$$
\left.
\begin{aligned}
U_E &= \frac{1}{2} \varepsilon E^2 \\
U_B &= \frac{1}{2\mu} B^2
\end{aligned}
\right\} \Rightarrow U_E = U_B; \ U_T = \varepsilon E^2
\tag{4.1.34}
$$

where U_T is the total energy.

4.1.5 Phase and Group Velocity

The interference of two electromagnetic waves E_1 and E_2 is

$$E_T = E_1 + E_2 = E_0 e^{i(k_1 x \omega_1 t)} + E_0 e^{i(k_2 x \omega_2 t)} = Ir \cdot Ca$$
$$Ir = 2E_0 \cos(k_l x - \omega_l t)$$
$$Ca = \cos(k_{\text{ave}} x - \omega_{\text{ave}} t)$$
$$\omega_{\text{ave}} = (\omega_1 + \omega_2)/2 \qquad\qquad (4.1.35)$$
$$\omega_l = (\omega_1 - \omega_2)/2;$$
$$k_{\text{ave}} = (k_1 + k_2)/2$$
$$k_l = (k_1 - k_2)/2$$

where Ir is the envelope or the irradiance which is slowly varying in amplitude and Ca is the carrier wave since it is the rapidly oscillating wave. The group velocity is the velocity of Ir. It is defined as

$$v_g \equiv \Delta\omega/\Delta k \qquad\qquad (4.1.36)$$

and more generally

$$v_g \equiv d\omega/dk. \qquad\qquad (4.1.37)$$

The carrier wave or the phase front propagates at the phase velocity. The phase velocity comes from the rapidly varying part:

$$v_\varphi \equiv \omega_{\text{ave}}/k_{ave}. \qquad\qquad (4.1.38)$$

Typically ω remains the same when transitioning from one medium to another, k however would change in this transition. Therefore,

$$\left.\begin{array}{l} v_g \equiv (dk/d\omega)^{-1} \\ k = n\omega/c = 2\pi/\lambda \end{array}\right\} \Rightarrow v_g = v_\varphi \bigg/ \left(1 + \frac{\omega}{n}\frac{dn}{d\omega}\right) = v_\varphi \bigg/ \left(1 - \frac{\lambda}{n}\frac{dn}{d\lambda}\right).$$

$$(4.1.39)$$

In a vacuum where the index of refraction n is independent of the frequency, the group velocity and the phase velocity are equal. However, generally the group velocity is slower than the phase velocity since n increases with ω.

4.1.6 Polarization

In the solution to Maxwell's equations:

$$E_x = E_0 \cos(\omega t - kz)$$
$$B_y = \frac{E_0}{c} \cos(\omega t - kz). \qquad (4.1.40)$$

The electric field varies only in the x-direction. The plane of polarization can be oriented in any direction. An electromagnetic (EM) wave can be represented as a composite of two orthogonal waves:

$$E_x = E_{0x} \cos(\omega t - kz - \varphi_x)$$
$$E_y = E_{0y} \cos\left(\omega t - kz - \varphi_y\right) \qquad (4.1.41)$$

where φ_x and φ_y are phases that specify the relative timing of the field oscillation. Figure 4.7 shows various EM polarization that depends on φ_x, φ_y, E_{0x}, and E_{0y}.

The polarization angle and amplitude of the EM wave are given by

$$\theta_p = \tan^{-1}\left(E_{0x}/E_{0y}\right)$$
$$A_p = \left(E_{0x}^2 + E_{0y}^2\right)^{1/2}. \qquad (4.1.42)$$

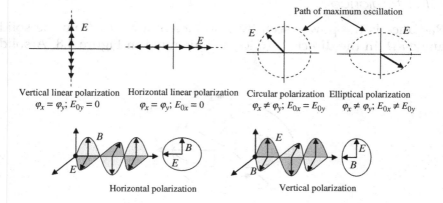

Path of maximum oscillation

Vertical linear polarization
$\varphi_x = \varphi_y; E_{0y} = 0$

Horizontal linear polarization
$\varphi_x = \varphi_y; E_{0x} = 0$

Circular polarization
$\varphi_x \neq \varphi_y; E_{0x} = E_{0y}$

Elliptical polarization
$\varphi_x \neq \varphi_y; E_{0x} \neq E_{0y}$

Horizontal polarization

Vertical polarization

Figure 4.7 Typical EM polarization

Table 4.2 Polarization types (max here refers to the maximum trace of the oscillations)

	S_0	S_1	S_2	S_3	E_{0x}	E_{0y}	φ_x	φy
x-polarized	1	1	0	0	0	X	φ_y	φ_x
y-polarized	1	-1	0	0	X	0	φ_y	φ_x
45 degree linear	1	0	±1	0	E_{oy}	E_{ox}	φ_y	φ_x
Circularly polarized (max)	1	0	0	±1	E_{oy}	E_{ox}	φ_x	φ_y
Elliptical polarized (max)	1	0.6	0	0.8	E_{ox}	E_{oy}	φ_x	φ_y

The general state of polarization is commonly specified using the Stokes vector:

$$
\begin{aligned}
S_0 &= \langle E_{0x}^2 \rangle + \langle E_{0y}^2 \rangle \\
S_1 &= \langle E_{0x}^2 \rangle - \langle E_{0y}^2 \rangle \\
S_2 &= \langle 2E_{0x}E_{0y} \cos(\varphi_y - \varphi_x) \rangle \\
S_3 &= \langle 2E_{0x}E_{0y} \sin(\varphi_y - \varphi_x) \rangle,
\end{aligned}
\tag{4.1.43}
$$

where S_0 is the total amplitude, S_1 is the amplitude of linearly polarized light, S_2 is the amplitude of right-circular polarization, and S_3 is the amplitude of left-circular polarization. The $\langle \ldots \rangle$ denotes a time average. Table 4.2 illustrates the various values of S and E for a given polarization.

4.1.7. Radiance

Radiance is the power seen by a surface of area dA from the solid angle $d\Omega$, in the direction (θ, φ) as illustrated in Figure 4.8. A solid

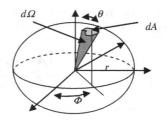

Figure 4.8 Solid angle

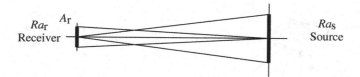

Figure 4.9 Source and receiver

angle Ω is used to represent a range of directions. It is defined as the area on the surface of a sphere divided by the square of the radius of the sphere:

$$d\Omega = dA/r^2$$
$$d\Omega = \sin(\theta)d\theta d\phi. \qquad (4.1.44)$$

Units used in plane angles are radians, but the solid angle Ω is in steradians and there are 4π steradians in a sphere (Figure 4.8).

The radiance is defined as

$$Ra = \frac{d\Phi}{dAd\Omega}. \qquad (4.1.45)$$

In Figure 4.9 the surface is perpendicular to the source. Φ is power or flux, and A is the area normal to the radiation. The radiance is invariant over a path in a vacuum.

$$\left. \begin{array}{c} \Omega_r = \dfrac{A_s}{r^2} \\[2mm] \Omega_s = \dfrac{A_r}{r^2} \end{array} \right\} \Rightarrow \left. \begin{array}{c} A_s\Omega_s = A_r\Omega_r \\[2mm] \Phi_s = \Phi_r \end{array} \right\} \Rightarrow Ra_r = \frac{\Phi_r}{A_r\Omega_r} = \frac{\Phi_s}{A_s\Omega_s} = Ra_s. \quad (4.1.46)$$

For spacebased sensors there are many EM sources that a sensor would see. The total energy seen is the sum of all the sources. Figure 4.10 illustrates a typical EM source and the interaction of the radiation with media such as atmospheric layers, ground and target, before reaching the spacebased sensor. The total radiant energy measured at the spacecraft L_{sc} as illustrated in Figure 4.10 is

$$L_{sc}(\lambda) = L_{surf} + L_{up} + L_{solar} + L_{atm} + L_{cloud} \qquad (4.1.47)$$

where L_{surf} is the surface emissions, L_{up} is the upwelled subsurface radiation, L_{solar} is the solar emissions which encompass surface

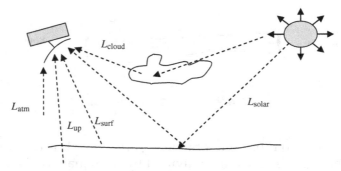

Figure 4.10 Typical sources of radiant energy received by a spacecraft

reflection and atmospheric scattering, L_{atm} is the atmospheric emissions, including direct emissions, surface reflections and atmospheric scattering, and L_{cloud} is the emission from clouds that include direct emissions and surface reflections.[21]

4.1.8 Irradiance

Irradiance, I often called intensity, describes the radiant power per unit area on a surface. The incident direction is not specified and is independent of the orientation. The surface is of a fixed size and has some orientation

$$I = \int_{\substack{\text{upward} \\ \text{hemisphere}}} L\cos(\theta)d\Omega = \int_{\theta=0}^{\pi/2} \int_{\varphi=0}^{2\pi} L\cos(\theta)\sin(\theta)d\theta d\varphi$$

$$I = \int_{\substack{\text{downward} \\ \text{hemisphere}}} L\cos(\theta)d\Omega = \int_{\theta=\pi/2}^{\pi} \int_{\varphi=0}^{2\pi} L\cos(\theta)\sin(\theta)d\theta d\varphi. \tag{4.1.48}$$

A light wave's average power per unit area is the irradiance:

$$I(r,\tau) = \langle S(r,\tau) \rangle_T = \frac{1}{T} \int_{\tau-T/2}^{\tau+T/2} S(r,t)dt. \tag{4.1.49}$$

Substituting a light wave into the expression for the Poynting vector,

$$S = c^2 \varepsilon E \times B = c^2 \varepsilon E_0 \times B_0 \cos^2(k \cdot r - \omega t - \theta) \Rightarrow I(r, \tau) = \frac{c^2 \varepsilon}{2} E_0 \times B_0.$$

$$(4.1.50)$$

Since the electric and magnetic fields are perpendicular and $B_0 = E_0/c$,

$$I = \frac{c\varepsilon}{2} |E_0|^2. \qquad (4.1.51)$$

Using the propagation properties in a medium:

$$I = \frac{n\varepsilon_0 c_0}{2} |E_0|^2$$

$$n = \frac{c_0}{c}. \qquad (4.1.52)$$

4.1.9 Interference

EM interference is the process of combining two or more EM waves. Using equation (4.1.32), the electric field wave is described by

$$E = E_0 \exp(i(k \cdot r - \omega t)). \qquad (4.1.53)$$

Combining two waves, the total irradiance is

$$I = \frac{c\varepsilon E_T \cdot E_T^*}{2} = \frac{c\varepsilon}{2} (E_1 + E_2) \cdot (E_1 + E_2)^*$$

$$I = I_1 + I_2 + c\varepsilon \Re(E_1 \cdot E_2^*) \qquad (4.1.54)$$

where the * refers to the complex conjugate. If the two waves have different polarizations the real part of the cross terms is null. If the two waves have the same polarization such as the combination of the same beam but one is delayed, fringes are obtained. Therefore, equation (4.1.54) becomes

$$I = 2I_0 + c\varepsilon \Re(E_0 \exp(i(\omega t - k_1 \cdot r)) E_0 \exp(i(k_2 \cdot r - \omega t)))$$

$$I = 2I_0(1 + \cos(k_2 \cdot r - k_1 \cdot r)). \qquad (4.1.55)$$

$$I = 2I_0(1 + \cos(2kx \sin(\theta))). \qquad (4.1.56)$$

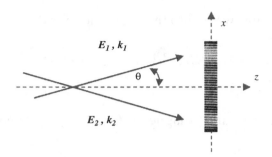

Figure 4.11   Interference of two EM waves

The maxima and minima of this equation correspond to fringes as illustrated in Figure 4.11, where the spacing between the fringes Δx is

$$I = 0 \Rightarrow 2kx\sin(\theta) = (2m+1)\pi \Rightarrow x = (2m+1)\lambda/(4\sin(\theta))$$
$$\Delta x = \lambda/(2\sin(\theta)).$$

$$(4.1.57)$$

The spacing between fringes increases as θ decreases. Figure 4.12 shows a spacebased sensor that uses optical interferometry. The total irradiance of the two fields at the two apertures is

$$I \propto 2 + 2\cos(kd)$$
$$d = b\sin(\theta) + d_1 - d_2$$

$$(4.1.58)$$

where d is the delay between the two interfering EM, b is the baseline (distance between the two collectors), θ is the angle between the base and the phase front, and d_1 and d_2 are the optical delay line 1 and line 2.

The phase front from a distant light source is coherent, and any subsequent interferences generate fringes. The coherence length l_c which is the distance over which the wavefront remains flat is

$$l_c = c \cdot \tau_c = c/\Delta v$$

$$(4.1.59)$$

where τ_c is the coherence time; i.e. the time over which the wavefront remains equally spaced, and Δv is the spectral bandwidth of the

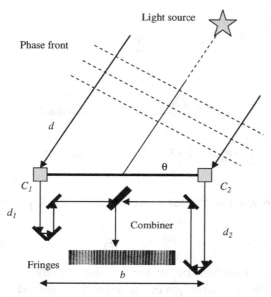

Figure 4.12 Optical interferometer geometry

incident source. Large bandwidth EM waves are incoherent. The spatial coherence depends on the emitter size and its distance from the target. The Von Cittert–Zernike theorem states that the spatial coherence area A_c is given by

$$A_c = D^2\lambda^2/\pi d^2 \tag{4.1.60}$$

where d is the diameter of the light source and D is the distance away.[10] Since stars are very distant, the incoming radiation is coherent.

4.1.10 Diffraction

Diffraction results from the incident radiation on an optic aperture. The outcome is a sequence of fringes on the edges of the field. This effect limits the ability to distinguish between two closely spaced objects such as stars. Typically, 84% of the light is concentrated in the central spot as shown in Figure 4.13. The rest falls in surrounding rings.

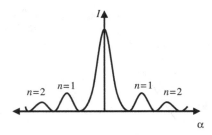

Figure 4.13 Diffraction pattern

For a telescope with diameter D, minima occur at positions given by

$$\sin \alpha_n = \frac{m_n \lambda}{D} \qquad (4.1.61)$$

where n is the number or order of the minimum and m is the numerical factor for any given n which is found by integrating over the light pattern. For small α:

$$\alpha_n = \frac{m_n \lambda}{D}. \qquad (4.1.62)$$

For $n = 1, 2$ and 3, m_n is respectively 1.22, 2.23 and 3.24. 84% of the light contained within the diameter defined by the first minimum is called the Airy disk, where

$$d_{\text{Airy}} = \frac{2.44 \lambda}{D} \qquad (4.1.63)$$

and this is used to define the resolution. In order to resolve two points' source, the closest peak of any other Airy disk needs to be further than its own first minimum (Figure 4.14).

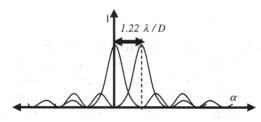

Figure 4.14 Separation between two sources

4.1.11 Black body Radiation

A black body at thermal equilibrium is an ideal absorber. It is also a perfect emitter: $\varepsilon_b = 1$. A black body emits radiation in all directions, at all wavelengths, at every temperature. The wavelength distribution is given by Planck's formula:

$$M_\lambda = \frac{2\pi c_1}{\lambda^5 (\exp(c_2/\lambda T) - 1)}$$
$$c_1 = hc^2$$
$$c_2 = hc/k$$

(4.1.64)

where M_λ is the spectral exitance or power emitted per unit area at wavelength λ, $c = 3 \times 10^8$ (in ms^{-1}) is the speed of light in a vacuum, $h = 6.63 \times 10^{-34}$ (in Js) is Planck's constant, T is the absolute temperature, λ is the wavelength and $k = 1.38 \times 10^{-23}$ (in JK^{-1}) is Boltzmann's constant. This equation is plotted in Figure 4.15 as a function of the wavelength for different temperatures.

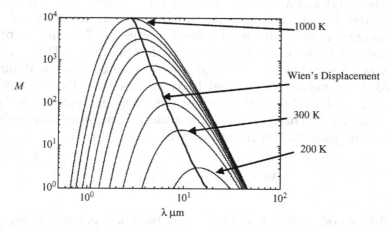

Figure 4.15 Spectral distribution of energy radiated from black-bodies of various temperatures[†]

[†]Figure 4.15 was generated using the following Matlab code. $c = 2.997e8; h = 6.625e - 34; k = 1.38e - 23; T0 = 200; i = 1e - 8;$ for $t = 0 : 8; T = T0 + 100^*t;$ lamda $= i : i/10 : 10000^*i; c1 = 2.^*pi.^*h.^* (c.^\wedge 2)./$ lamda.$^\wedge 5; c2 = h.^*c./(k.^*T.^*$ lamda $); M = 2.77e - 7.^*c1./(exp((c2) - 1));$ loglog(lamda./$10^\wedge (-6),M$);axis([.5 100 1 10^4]);hold on;end.

The Stefan–Boltzmann Law states that all matter at temperatures above absolute zero continually emits EM radiation. The higher the temperature the greater the power emitted and the 'bluer' the light emitted. In addition to the sun, terrestrial objects are also sources of radiation, though of a different magnitude and spectral composition than the sun. Black body radiation is emitted from a hot body. It results from a combination of spontaneous emission, stimulated emission and absorption occurring in a medium at a given temperature. Objects at high temperatures emit short wavelengths and objects at low temperatures emit long wavelengths. For instance, the sun at $11\,000°\,\mathrm{F}$ emits the energy primarily in the visible spectrum. A human body at $98.6°\,\mathrm{F}$ emits longer IR wavelengths.

The Stefan–Boltzmann Law is derived from the integration of Planck's Law with respect to wavelength:

$$\int M_\lambda d\lambda = M_{\text{total}} = \sigma T^4 \tag{4.1.65}$$

where σ is the Stefan–Boltzmann constant $5.6697 \times 10^{-8}\ \mathrm{Wm^{-2}K^{-4}}$, M is the total radiant exitance from the surface of a material in $\mathrm{Wm^{-2}}$, and T is the temperature in K of the emitting material. The total energy emitted from an object varies as T^4 and therefore increases rapidly with increases in temperature. This law is expressed for a black body like energy source, an ideal object; therefore the exitance of a real body is less than that of a black body. The emissivity ε is the ratio of the object exitance to the black body equivalent. The emissivity ε is a measure of how efficiently an object emits radiation as compared with a black body:

$$M = \varepsilon M_{\text{Black body}} = \varepsilon \sigma T^4. \tag{4.1.66}$$

For a black body $\varepsilon = 1$, and for a white body $\varepsilon = 0$. Since emissivity, like exitance, is a spectral quantity, therefore

$$M_\lambda = \varepsilon_\lambda M_{\text{Black body }\lambda}$$
$$M_{\Delta\lambda} = \int_{\lambda_1}^{\lambda_2} \varepsilon_\lambda M_{\text{Black body }\lambda} d\lambda. \tag{4.1.67}$$

Table 4.3 Typical emissivity values[14]

Material	Concrete	Sand	Soil	Vegetation	Water	Aluminum	Polished brass	Anodized brass
ε	0.95	0.9	0.92–0.94	0.88–0.94	0.98	0.55	0.03	0.61

Emissivity depends on the nature of the material, its temperature and its surface characteristics, such as roughness (Table 4.3). Typically, for metals, the emissivity is less than 0.5, but when the temperature is increased or the surface oxidizes, then the emissivity increases. The emissivity of polished metals can be as low as 0.01 which is very useful for IR optical systems.

The total exitance from a real object is

$$M_{det} = M_{emit} + M_{ref} = \varepsilon\sigma T^4 + \rho E \qquad (4.1.68)$$

where ρ is the reflectivity, M_{ref} is the reflected exitance. Using equation (4.1.64) for large λ, the exitance is described by the Rayleigh–Jeans Law:

$$M_\lambda \approx \frac{2\pi c_1}{\lambda^4 c_2/T}. \qquad (4.1.69)$$

Using equation (4.1.64) for small λ, the exitance is described by Wien's Law:

$$M_\lambda \approx \frac{2\pi c_1}{\lambda^5 \exp(c_2/\lambda T)}. \qquad (4.1.70)$$

Wien's Displacement Law determines the dominant wavelength at which a black body radiation curve reaches a maximum (Figure 4.15). If Planck's equation is differentiated with respect to wavelength and set equal to 0 to find the maxima of the emission curve, one arrives at Wien's Displacement Law:

$$\partial_\lambda M_\lambda = 0 \Rightarrow \lambda_{max} = \frac{2898}{T} \qquad (4.1.71)$$

where λ_{max} is the wavelength at which exitance is at a maximum. This means that very hot sources emit in the ultraviolet, and cold sources emit radio waves. As the temperature of an object gets hotter, λ gets shorter. For example, the temperature of the sun is 5700 K which corresponds to $\lambda_{sun} = 0.51\,\mu m$ the middle of the visible spectrum. For the human body at temperature 310 K, the $\lambda_{body} = 9.4\,\mu m$, which is in the TIR.

The emitted radiation by a target over a band of wavelengths is only a fraction of the total radiation defined as

$$F_{\Delta\lambda} = \frac{M_{\Delta\lambda}}{M_{total}} = \frac{1}{\sigma T^4}\int_{\lambda_1}^{\lambda_2} M_\lambda d\lambda. \qquad (4.1.72)$$

Radiometric temperature is mainly measured using one or two reference sources with a known temperature. The measurement is then conducted by alternating between the reference temperature T_r and the target temperature T_t. The relationship between the levels of exitance from the target and the reference source is

$$\frac{M_t}{M_r} = \left|\frac{F_t\sigma T_t^4}{F_r\sigma T_r^4}\right| \Rightarrow T_t = \left|\frac{M_t F_t}{M_r F_r} T_r^4\right|^{1/4} \qquad (4.1.73)$$

where M_t and M_r are the exitance from the target and the reference, and F_t and F_r, are the fractions of radiation received over wavelengths being sensed. The reference source can be an integral part of the sensor. The sensor will measure the area of interest and the reference alternately by using, for example, a rotating chopper or other mechanical mechanism.

4.2 Interaction with Matter

The interaction between incident electromagnetic energy and a given surface is summarized by three fundamental energies: reflection, absorption and transmission. These three energy forms are related, using the conservation of energy principle, by

$$E_i(\lambda) = E_r(\lambda) + E_a(\lambda) + E_t(\lambda) \qquad (4.2.1)$$

where E_i is the incident energy, E_r is the reflected energy, E_a is the absorbed energy, and E_t is the transmitted energy. For molecules, each state has many levels that are related to nuclear spin, vibration and rotation. Atoms have simpler systems and hence simpler

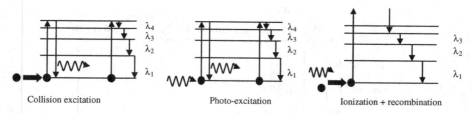

Figure 4.16   Discrete atomic emission

spectra. In atoms, electrons have only certain discrete energy levels to transition between. They are accompanied by the emission or absorption of photons. The absorbed or emitted photon energy corresponds to the energy lost or gained in the transition. Atoms spontaneously decay to the ground state after a time. If an electron gains more energy than the ionization potential then it is no longer bound to the atom. Only the lowest level, which is the ground state, is generally stable. Excited states have typical lifetimes of the order of $\sim 10^{-8}$ s. After this time they fall back to the most stable state and emit a photon in a random direction whose energy is equal to the difference between the two states.

An emission line is produced when an electron in an excited state falls by one or more levels. Levels are populated when electrons undergo a collision excitation, a photo-excitation or a recombination (Figure 4.16). The emitted wavelength depends on the energy of the excitation source

$$\lambda = hc/\Delta E. \tag{4.2.2}$$

Electrons can reach higher energy levels if they collide with other electrons, ions or atoms that have enough kinetic energy for this transition. Energy is radiated when electrons drop from higher levels to lower energy levels. Ionization occurs when a sufficiently high-energy photon interacts with an atom so that an electron leaves the atom. This electron could recombine with an ion and emit photons until it reaches its ground state.

4.2.1 Atmospheric Absorption

Atmospheric absorption results in the effective loss of energy at a given wavelength to atmospheric constituents. Solar radiation is mainly absorbed by water vapor, carbon dioxide and ozone. Between

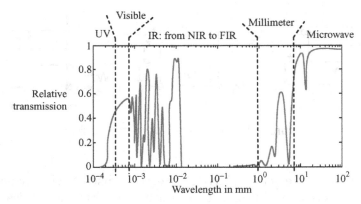

Figure 4.17 Typical EM wave attenuation by the atmosphere[14]

the sensor and the target, the emitted or reflected radiation passes through the intervening media and interacts with the constituents of that media. Typically this attenuation follows Lambert's Law

$$I = I_0 e^{-kd} \qquad (4.2.3)$$

where I_o is the radiation incident on the attenuating medium, k is the extinction coefficient and d is the path length. This equation shows that the longer the path, the higher the absorption and the lower the transmission.

Figure 4.17 shows the presence of atmospheric windows which are areas that have low attenuation for a given wavelength. In the IR region, water and ozone are some of the primary elements which cause absorption, as presented in Figure 4.18. Therefore IR detection has to operate in a specific window. For instance, if the sensor needs to be designed to measure ozone (O_3), then the wavelength for which this sensor is designed should use a wavelength of $\sim$10 μm since that is where a large peak occurs, as illustrated in Figure 4.18. For microwaves, Figure 4.17 shows that the atmosphere is virtually transparent.

4.2.2 Reflectance

The reflectance of a surface depends on the incident wavelength and the structure of the surface. If the wavelengths are much smaller than

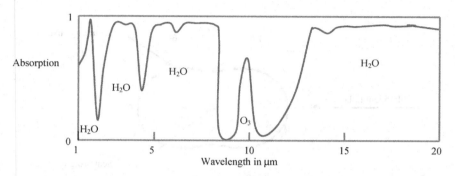

Figure 4.18 Typical atmospheric window in the IR region[14]

the surface variations or the particle sizes, diffuse reflection will dominate. Figure 4.19 illustrates various types of reflections that are surface structure and wavelength dependent. Specular or mirror-like reflection will dominate when the wavelength is much larger than the surface structure. A Lambertian surface would reflect isotropically the incident electromagnetic wave. This type is often used as an ideal surface.

4.2.3 Scattering

Atmospheric scattering results from the interaction with the atmosphere, the path length that the radiation travels from the target to the sensor and the radiation wavelength of interest. There are many types of scattering that are intimately related to the medium, such as particle size. The following paragraphs describe some of these scattering mechanisms.

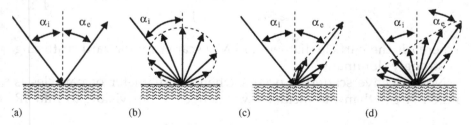

Figure 4.19 Surface types: (a) specular, (b) Lambertian, (c) quasi-specular, and (d) quasi-Lambertian[6]

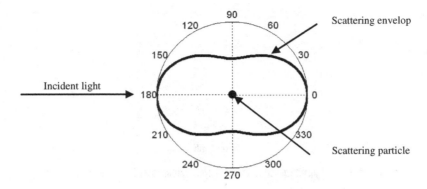

Figure 4.20 Rayleigh scattering envelope[†]

Rayleigh scattering occurs when the wavelength of the incident EM radiation is much larger than the particle size. The intensity of the Rayleigh scatter is

$$I_{scat} = I_0 \frac{8\pi^4 n\alpha^2}{\lambda^4 R^2} \left(1 + \cos^2(\theta)\right) \propto \frac{1}{\lambda^4} \qquad (4.2.4)$$

where n is the number of particles present in the medium, α is the polarizability of particles or molecules and R is the distance from the scatterer. This equation shows that the shorter the radiation wavelength the more scattering results.[19] Rayleigh scattering is dominant for the scattering of particles up to about a tenth of the wavelength of the light (Figure 4.20).

Mie scattering is dominant when the wavelength is of the same order as the particle diameter. For example, haze and smog cause Mie scattering. Typically, the scattering I is

$$r \approx \lambda \Rightarrow I \propto 1/\lambda^2$$
$$r \approx 3\lambda/2 \Rightarrow I \propto 1/\lambda \qquad (4.2.5)$$

where r is the particle diameter. In Mie scattering forward scattering is strongly dominant.[19]

Non-selective scattering occurs when the diameter of particles is much larger than the incident wavelength. From visible to mid IR

[†] Figure 4.20 was generated using the following Matlab code. theta=0 : 0.1 : 6; y=1+ cos(theta). ^2;polar (theta},y).

water droplets cause this type of scattering. Non-selective scatter comes from the indifference of the mechanism to a wavelength range. Absorption in atmospheric windows results from this effect.

Raman scattering is an inelastic scattering where the scattered photon's frequency is shifted after interaction between incident photons and the molecules. This scattering depends on the polarizability of the molecules. Indeed, vibrational modes of molecules can be excited by the incident photons. The energy of the scattered photons changes. Spectral analysis of the scattered light shows Stokes lines when the scattered frequency is lower than the incident. Anti-Stokes lines are observed when the scattered frequency is higher than the incident. This is very useful for particle identification in remote sensing.

Compton scattering results from the interaction between light and charged particles. The scattered light or photons will have less energy and less momentum after this interaction, and so should have a larger wavelength according to the formula:

$$\Delta\lambda = \lambda_{scattered} - \lambda_{incident} = h(1 - \cos(\theta))/mc \Rightarrow \Delta\lambda_{max} = 2h/mc$$

$$(4.2.6)$$

where h is Planck's constant and $\Delta\lambda_{max}$ is the maximum wavelength change from charged particle scattering. This value is significant for X-ray and gamma-ray radiation.

4.3 Optics

For optic-based sensors, the incoming light is filtered, collimated, spread and focused onto sensing elements. This sequence relies on a set of mirrors, lenses and other optical components. In this section an overview of optical physics is presented.

4.3.1 Refraction/Reflection

At an interface (Figure 4.21), and for perpendicularly polarized EM waves, the relationship between the incident, the reflected and the transmitted EM is

$$E_{0i} = E_{0r} + E_{0t}$$
$$B_{0i}\cos(\theta_i) = B_{0r}\cos(\theta_r) - B_{0t}\cos(\theta_t).$$

$$(4.3.1)$$

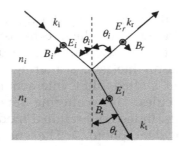

Figure 4.21 Perpendicularly polarized EM wave[8]

Using the relationship between E and B:

$$E = \frac{c_0}{n} B, \theta_i = \theta_r \Rightarrow \begin{cases} r_\perp = \dfrac{E_{0r}}{E_{0i}} = \dfrac{n_i \cos(\theta_i) - n_t \cos(\theta_t)}{n_i \cos(\theta_i) + n_t \cos(\theta_t)} \\[4mm] t_\perp = \dfrac{E_{0t}}{E_{0i}} = \dfrac{2 n_i \cos(\theta_i)}{n_i \cos(\theta_i) + n_t \cos(\theta_t)} \end{cases} ; \qquad (4.3.2)$$

where r and t are the reflection and transmission coefficients for perpendicularly polarized EM waves. These equations are called the Fresnel equations for perpendicular polarized EM waves (Figure 4.21).
For parallel polarized electromagnetic waves:

$$B_{0i} - B_{0r} = B_{0t}; E$$
$$\cos(\theta_i) + E_{0r}\cos(\theta_r) = E_{0t}\cos(\theta_t). \qquad (4.3.3)$$

Therefore the parallel transmission and reflection coefficients are

$$\begin{aligned} r_\| &= \frac{E_{0r}}{E_{0i}} = \frac{n_i \cos(\theta_t) - n_t \cos(\theta_i)}{n_i \cos(\theta_i) + n_i \cos(\theta_i)} \\[3mm] t_\| &= \frac{E_{0t}}{E_{0i}} = \frac{2 n_i \cos(\theta_i)}{n_t \cos(\theta_i) + n_i \cos(\theta_t)}. \end{aligned} \qquad (4.3.4)$$

These equations are called the Fresnel equations for parallel polarized EM waves. For common vacuum/glass interfaces, the indices of refraction are $n_{\text{vacuum}} = 1 < n_g \approx 1.5$. Note that total reflection occurs

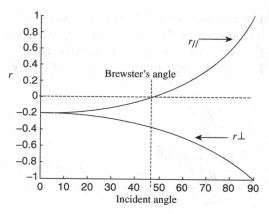

Figure 4.22 Incident angle effect on reflection coefficient for vacuum to glass and glass to vacuum[†]

at $\theta = 9\overset{\circ}{0}$ for both polarizations as presented in Figure 4.22. Zero reflection for parallel polarization occurs at 56.3°. This angle is called Brewster's angle (Figure 4.22). Brewster's angle is widely used in applications where the design requires zero reflection. This occurs for most laser applications where the lasing medium (Figure 4.23), which could be crystalline, amorphous, liquid, gaseous or plasma within the resonating cavity, is subject to various light passes. Each pass or transition within this lasing medium requires a zero reflec-

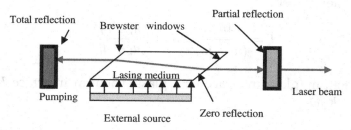

Figure 4.23 Optical resonator uses Brewster's angle in laser applications to avoid reflective losses

[†] Figure 4.22 was generated using the following Matlab code. $ni=1; nt=1.5; qt=0; i=0 : 0.01 : 0.25;$ $t=i*2*180;$ $ru=ni*\cos(2*pi*i) - nt*\cos(qt);$ $rd=ni*\cos(2*pi*i) + nt*\cos(qt);$ $r1=ru./rd; ru=ni*\cos$ $-nt*\cos(2*pi*i); rd=nt*\cos(2*pi*i) + ni*\cos(qt); r2=ru./rd;$ hold on; $plot(t,r2,'k');$ hold on; plot $(t,r1,'k').$

tion for maximum performance. Therefore Brewster windows, which are windows that are set at Brewster's angle, are used (Figure 4.23).

Looking at the transmitted power versus reflected power, the transmittance T and the reflectance R are used. They are defined as follows:

$$\left. \begin{array}{l} T = \dfrac{I_t A_t}{I_i A_i} \\[2mm] I = \frac{1}{2} n \varepsilon_0 c_0 |E_0|^2 \\[2mm] t^2 = \dfrac{|E_{0t}|^2}{|E_{0i}|^2} \\[2mm] \dfrac{A_t}{A_i} = \dfrac{d_t^2}{d_i^2} = \dfrac{\cos(\theta_t)}{\cos(\theta_i)} \end{array} \right\} \Rightarrow T = \dfrac{n_t \cos(\theta_t)}{n_i \cos(\theta_i)} t^2 \qquad (4.3.5)$$

where T is the transmittance or the transmissivity, which is the transmitted power divided by the incident power, A is the area, and d is the beam diameter, as illustrated in Figure 4.24.

$$\left. \begin{array}{l} \theta_i = \theta_t \\[2mm] n_i = n_r \\[2mm] R = \dfrac{I_r A_r}{I_i A_i} = \dfrac{n_r |E_{0r}|^2}{n_i |E_{0i}|^2} \dfrac{A_r}{A_i} \end{array} \right\} \Rightarrow R = r^2 \qquad (4.3.6)$$

where R is the reflectance or the reflectivity, which is the reflected power divided by the incident power. Reflection at normal incidence occurs when $\theta_i = 0$,

$$T = \frac{4 n_i n_t}{(n_t + n_i)^2}; R = \left(\frac{n_i - n_t}{n_t + n_i} \right)^2. \qquad (4.3.7)$$

For a vacuum/glass interface or a glass/vacuum interface $R = 4\%$ and $T = 96\%$.

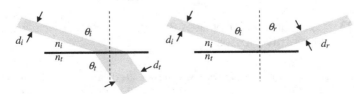

Figure 4.24   Incident, transmitted and reflected beam

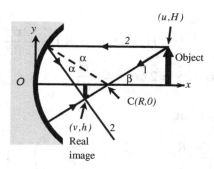

Figure 4.25   Object and its image with a concave mirror

4.3.2 Concave Mirror

Concave mirrors are used to gather light at their focus. The larger this reflecting surface, the more light is gathered. For spacebased sensors, the extent of the reflector is limited because of weight and area since it has to fit on the launch vehicle. In this section, basic optics geometry is presented for mirrors followed by lenses.

In Figure 4.25, C is the centre of curvature of the concave mirror, R is its radius, (u, H) is the location of the object. Using ray tracing of rays 1 and 2 from object to image, the following relation is derived

$$\frac{h}{v} = \frac{H}{u}$$

$$\frac{h}{R - v} = \tan \beta = \frac{H}{u - R}$$

$$\frac{R - v}{v} = \frac{u - R}{u} \qquad (4.3.8)$$

$$\frac{1}{u} + \frac{1}{v} = \frac{2}{R}$$

Assuming that all points are close to the optic axis and therefore the angles are small, which constitutes the paraxial approximation, ray 1 follows:

$$y = (x - R) \tan \beta = \frac{H(x - R)}{u - R}. \qquad (4.3.9)$$

For ray 2:

$$y \approx H - x\tan 2\alpha \approx H - \frac{2xH}{R}$$

$$\tan 2\alpha \approx 2\tan\alpha \approx \frac{2H}{R}. \qquad (4.3.10)$$

Rays 1and 2 intersect at

$$\frac{x-R}{u-R} = 1 - \frac{2x}{R} \Rightarrow x\left(\frac{2}{R}-\frac{1}{u}\right) = 1$$

$$\frac{1}{u}+\frac{1}{v} = \frac{2}{R} = \frac{1}{f} \Rightarrow \begin{cases} x = v \\ y = h \end{cases}. \qquad (4.3.11)$$

Within paraxial approximation all rays from (u,H) pass through (v,h) forming an inverted real image where f is the focal length of the mirror, u is the object and v is the image distance. The lateral magnification is

$$M_T = \frac{h}{H} = -\frac{v}{u} \qquad (4.3.12)$$

The diagrams in Figure 4.26 illustrate typical cases of spherical surface reflection. The image of the object depends on u, v and M_T.

4.3.3 Lenses

There are several types of lens that are used to collimate or expand incoming light to a sensing element. Figure 4.27 shows typical lenses

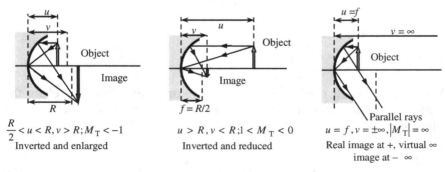

Figure 4.26 Reflection: spherical surface, typical cases[8]

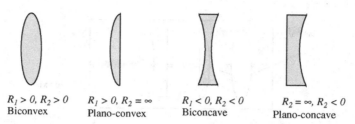

$R_1 > 0, R_2 > 0$ $R_1 > 0, R_2 = \infty$ $R_1 < 0, R_2 < 0$ $R_2 = \infty, R_2 < 0$
Biconvex   Plano-convex   Biconcave   Plano-concave

Figure 4.27 Typical lenses – R_1 and R_2 are the radius of curvature of side 1 and side 2, respectively

that could be used in optic systems. In general, a lens surface is a portion of a sphere. R is the radius of that sphere. Converging lenses have positive focal lengths (Figure 4.28). If $u > f$ the lens forms real images of real objects with $v > 0$. If $0 < u <$ the lens forms virtual images of real objects, e.g. a magnifying glass. When rays are parallel to the optic axis, converging lenses focus rays to a focal point, diverging lenses diverge rays from a focal point. Rays passing through a focus are refracted by the lens to emerge parallel to the axis.

The relationship between u, v and f for a thin lens is

$$D = \frac{1}{f} = \frac{1}{u} + \frac{1}{v} = (n_1 - 1)\left(\frac{1}{R_1} - \frac{1}{R_2}\right) \qquad (4.3.13)$$

where n_1, R_1, R_2, and D are the index of refraction of the lens, radius of the incident surface, the radius of the transmitting surface and the power of the lens respectively.

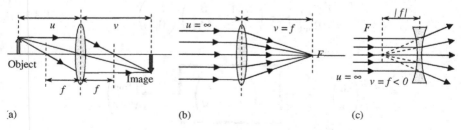

(a)   (b)   (c)

Figure 4.28 (a) and (b) Converging lens diagrams; (c) diverging lens diagram

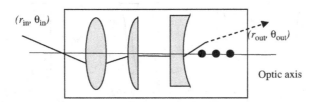

Figure 4.29 Lens combination

4.3.4 Lens Combinations

Optical systems, where a combination of lenses, reflectors, beam splitters and filters are present between the target and the sensor, could be analyzed using the transfer function principle used extensively for electronics.

Figure 4.29 shows that as a light ray propagates through a series of lenses the distance r from the optic axis to the light ray changes as well as the slope θ. The optical system could be viewed as a box (Figure 4.30) that has a transformation matrix that uses the input light ray parameters to derive where the light ray will emerge from these optics. A 2×2 matrix could be defined for all optical components. The system matrix is derived by the succession of these matrices. The effect on a ray is determined by multiplying its ray vector R_v.

In Figure 4.30 AM is the angular magnification, SM is the spatial magnification, r_{in} and θ_{in} are the position and slope upon entering, r_{out} and θ_{out} are the position and slope after propagating from $x = 0$ to x, and T is a 2×2 transformation matrix that describes the optic system. For small displacements and angles, one can derive that the output R_v is

$$\begin{pmatrix} r_{out} \\ \theta_{out} \end{pmatrix} = \begin{pmatrix} SM & B \\ C & AM \end{pmatrix} \begin{pmatrix} r_{in} \\ \theta_{in} \end{pmatrix}. \qquad (4.3.14)$$

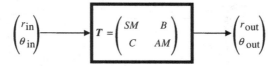

Figure 4.30 Incoming EM radiation on the front end of the optic is transformed to reach the sensor

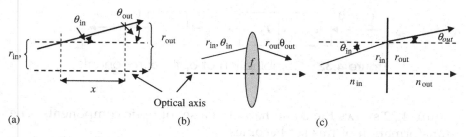

Figure 4.31 Ray (a) in a free space, (b) at a lens and (c) at an interface

Note the output ray vector of the interfaces illustrated in Figure 4.31.

The transformation matrix T_{fs} for a ray in a free space as presented in Figure 4.31(a) is

$$\begin{cases} r_{out} = r_{in} + x\theta_{in} \\ \theta_{out} = \theta_{in} \end{cases} \Rightarrow \begin{cases} \begin{pmatrix} r_{out} \\ \theta_{out} \end{pmatrix} = T_{fs} \begin{pmatrix} r_{in} \\ \theta_{in} \end{pmatrix} \\ T_{fs} = \begin{pmatrix} 1 & x \\ 0 & 1 \end{pmatrix} \end{cases}. \qquad (4.3.15)$$

The transformation matrix of a thin lens T_{tl} as presented in Figure 4.31(b), where only the ray slope changes, is

$$\begin{cases} r_{out} = r_{in} \\ \theta_{out} = -\frac{1}{f} r_{in} + \theta_{in} \end{cases} \Rightarrow \begin{cases} \begin{pmatrix} r_{out} \\ \theta_{out} \end{pmatrix} = T_{tl} \begin{pmatrix} r_{in} \\ \theta_{in} \end{pmatrix} \\ T_{tl} = \begin{pmatrix} 1 & 0 \\ -1/f & 1 \end{pmatrix} \end{cases}. \qquad (4.3.16)$$

At the interface where the slope changes based on Snell's law, the transformation matrix T_{int} as illustrated in Figure 4.31(c) is

$$\begin{cases} r_{out} = r_{in} \\ n_{out} \sin(\theta_{out}) = n_{in} \sin(\theta_{in}) \end{cases} \xrightarrow{\text{small angles}} \begin{cases} \begin{pmatrix} r_{out} \\ \theta_{out} \end{pmatrix} = T_{int} \begin{pmatrix} r_{in} \\ \theta_{in} \end{pmatrix} \\ T_{int} = \begin{pmatrix} 1 & 0 \\ 0 & n_{in}/n_{out} \end{pmatrix} \end{cases}. \qquad (4.3.17)$$

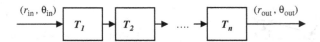

Figure 4.32 Combination of optic components

Figure 4.32 shows the combination of a set of optic components. The transformation of this set becomes

$$\begin{pmatrix} r_{\text{out}} \\ \theta_{\text{out}} \end{pmatrix} = T_n \dots T_2 T_1 \begin{pmatrix} r_{\text{in}} \\ \theta_{\text{in}} \end{pmatrix}. \qquad (4.3.18)$$

Note the transformation matrix for the lens shown in Figure 4.33 is

$$T = \begin{pmatrix} 1 & x_i \\ 0 & 1 \end{pmatrix} \begin{pmatrix} 1 & 0 \\ -1/f & 1 \end{pmatrix} \begin{pmatrix} 1 & x_o \\ 0 & 1 \end{pmatrix} = \begin{pmatrix} 1 - x_i/f & x_i + x_o - x_i x_o/f \\ -1/f & 1 - x_o/f \end{pmatrix}.$$
$$(4.3.19)$$

For $x_i = x_o = 0$, $T = T_{tl}$.

4.3.5 Aberrations

The refractive index of glass, and thus the focal length of a lens, is a function of wavelength. Since, for example, a single lens cannot focus red and blue light to the same point, there is chromatic aberration (Figure 4.34(a)). Indeed, different wavelengths have different speeds and therefore different refractive indices for the same material. In order to reduce chromatic aberration, two or more lenses made from different materials are used instead of a single material. Also scaling individual wavelengths independently can minimize the residual chromatic aberration.

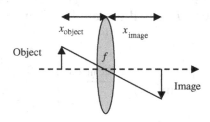

Figure 4.33 Lens example

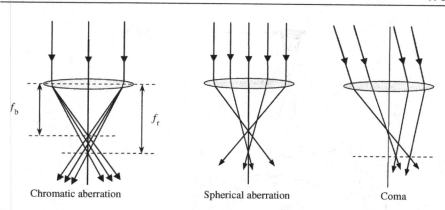

Chromatic aberration   Spherical aberration   Coma

Figure 4.34 Example of (a) chromatic aberration, (b) spherical aberration and (c) coma

Spherical aberrations occur when light rays which are parallel to the optical axis of a lens or spherical mirror, but which lie at different distances from the axis, are brought to different foci (Figure 4.34(b)). This results in the blurring of the image. Spherical aberration is aggravated by peripheral rays, by spherical mirrors and lens orientation, as well as by decreasing the focal length.

Coma is seen when the focus of off-axis light depends on the path it takes through the lens or where it reaches the mirror (Figure 4.34(c)). Coma occurs when parallel peripheral rays having a non-null angle with respect to the optic axis cross the focal plane at many points. This results in changes in magnification with aperture. The coma may point towards the axis (called positive coma), or away from it (called negative coma). Spherical mirrors do not suffer from coma because the mirrors always present the same geometry to the point source, irrespective of the off-axis angle. However, parabolic mirrors, which do not have spherical aberration, do suffer from coma, so are only effective in a narrow field around the optical axis.

4.3.6 Optical Resolution

The resolution of a sensor is the smallest difference between two units of measurements. Spatial resolution is the smallest angular or linear separation between two objects resolvable by the sensor. Typically, the spatial resolution should be less than half of the size of the smallest object of interest. Figure 4.35 illustrates a typical configuration of a spacebased optical sensor.

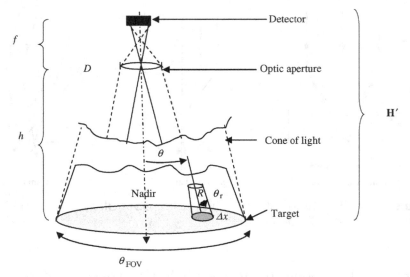

Figure 4.35 Spacebased optic (not to scale)[13]

The detector focal length f needed to record a target of radius R on the ground is

$$\frac{f}{h} = \frac{r_d}{R} = m \tag{4.3.20}$$

where h is the sensor altitude, r_d is the radius of the detector array and m is the magnification factor. At the diffraction limit, using equation (4.1.63), the required focal length f to give an image of diameter D_i for a point target is

$$f = D_i D / 2.44\lambda \tag{4.3.21}$$

where λ is the wavelength of light and D is the aperture diameter. The element defined by the angular resolution, which is the smallest angular separation or diffraction limited resolution, is

$$\theta_r = 1.22\frac{\lambda}{D}. \tag{4.3.22}$$

Therefore the ground resolution Δx is

$$\Delta x = 1.22\frac{\lambda h}{D\cos\theta}. \tag{4.3.23}$$

For imaging in astronomy, angular resolution is a more relevant measurement since targets are distant stars so

$$\theta_{FOV} = 2/\tan^{-1}\left(\frac{r_d}{f}\right) \qquad (4.3.24)$$

where r_d is the detector array radius. Large fields of view (FOV) are obtained for large detectors, whereas for long focal lengths the FOV is small

$$A = \pi R^2 = \pi h^2 \tan^2\left(\frac{\theta}{2}\right). \qquad (4.3.25)$$

The detector foot print, or the FOV, is the ground area a detector sees if the nadir is pointed straight down, so

$$D_g = H'\theta_{FOV} \qquad (4.3.26)$$

where D_g is the diameter of circular ground area viewed by the detector, H' is the height of the detector above the terrain, and θ_{FOV} is the angle (in radians) of the system's instantaneous field of view. These parameters are illustrated in Figure 4.35.

4.4 Scanning Mechanisms

There are various types of electro-optical imaging sensors as illustrated in Table 4.4. A combination of cross-track and along-track (see Figure 4.36), leads to the mapping of a surface of interest.

Because of the wider angular FOV, the cross-track approach is preferred for reconnaissance. Along-track is preferred to obtain

Table 4.4 Advantages and disadvantages of cross-track and along-track

	Cross-track	Along-track
Angular FOV	Wider	Narrower
Mechanical system	Complex	Simple
Optical system	Simple	Complex
Spectral range of detectors	Wider	Narrower
Dwell time	Shorter	Longer

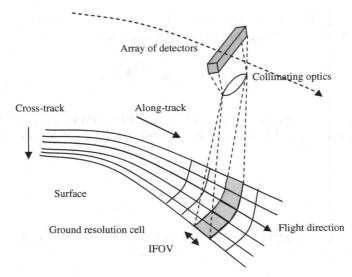

Figure 4.36 Typical pushbroom[19]

detailed spectral and spatial information. A problem associated with along-track is that calibration of thousands of detectors is required to achieve uniform sensitivity of the array. For along-track scanners the dwell time of ground resolution cell is only a function of velocity. Pushbroom scanners and especially whiskbroom scanners are typical methods that are used in spacebased sensors.

4.4.1 Linear Array: Pushbroom

Pushbroom sensors consist of a linear array of detectors where each detector is sensitive to a specific wavelength range. As the spacecraft moves forward each detector signal is sampled at regular intervals, depending on the speed of forward motion and on the altitude (see Figure 4.36). The spatial resolution is determined by the size of the raw detectors. Therefore the sensing element is generally designed to be very small, with a single array containing a few thousand detectors. Each spectral band has its own array of sensors that is placed in the focal plane of the scanner. Typically, the long axis of the array is oriented normal to the flight path.

The instantaneous field of view (IFOV) is the field of view of a single detecting element. It can be expressed in radians or steradians, but it is often used to refer to the pixel size in the sampled image. The image is created by light-sensitive detectors that produce electrical signals proportional to the brightness of the light energy. Each value recorded is a sample of a voltage level at a particular instant in time. The IFOV depends upon the altitude and resolution of the sensor. It is the smallest ground area for which digital measurement is collected. A variance of the pushbroom method is the hyperspectral area array which has many narrow spectral bands. In addition to the linear array a dispersing element is used.

4.4.2 Whiskbroom

While pushbroom sensors may have thousands of detectors per spectral band, scanning mirror sensors usually only have a few. A single detector can be made to view a strip of terrain by using a rotating mirror to direct its field of view. This is known as scanning. A whiskbroom is the assembly of a scanning mirror and a single discrete detector for each spectral band. A rotating mirror changes the angle of the incident light source and therefore the portion of the detected ground. This is the simplest configuration. The pixel width is a function of the mirror rotation rate and the IFOV. The pixel length is a function of the IFOV, sensor speed and sampling rate. The mirror angles the light across these multiple detectors instead of just one detector (see Figure 4.37). Sometimes whiskbrooms use a dispersing element such as a prism instead of wide-band filters to separate the incoming light into its component wavelengths. A

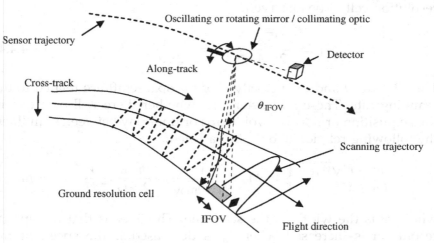

Figure 4.37 Typical whiskbroom[19]

rotating mirror and forward sensor movement create the spatial arrangement of pixels.

4.4.3 Scanner parameters

The length of time a detector sees a ground target is called the dwell time t_{dwell} which is the time required for a detector IFOV to sweep across a ground resolution cell. Longer dwell times yield stronger signals, so

$$t_{\text{dwell}} = \dot{s}/n \qquad (4.4.1)$$

where t_{dwell} is the dwell time, $\dot{s}$ is the scan rate per line and n is the number of cells per line. The detector dwell time is the time the detector must look at or dwell on a resolution element. In general the detector dwell time t_{dwell} per element is

$$t_{\text{dwell}} = kt_{\text{d}} \qquad (4.4.2)$$

where t_{d} is the minimum response time of the detector.

$$n_{\text{res}} = M \frac{2\pi}{\sqrt{\theta_{\text{IFOV}}}} \Rightarrow t_{\text{s}} = \frac{\sqrt{\theta_{\text{IFOV}}}}{2\pi M} \qquad (4.4.3)$$

where M is the rotation rate, n_{res} is the number of resolution elements scanned per second and t_{s} is the scan time per element, which is the time required to scan a single resolution element.[5] For each resolution cell to be observed:

$$t_{\text{s}} = \frac{\sqrt{\theta_{\text{IFOV}}}}{2\pi M} \geq kt_{\text{d}} \Rightarrow \frac{\sqrt{\theta_{\text{IFOV}}}}{2\pi k t_{\text{d}}} \geq M. \qquad (4.4.4)$$

The altitude H and the velocity V of the spacecraft are related to the scanning rate. These will determine whether there will be a gap in the acquisition or not. To avoid gaps or underlay between scan lines, the following relation should be satisfied:

$$\left. \begin{array}{l} w = H\sqrt{\theta_{\text{IFOV}}} \\ Mw > V \end{array} \right\} \Rightarrow M > \frac{V}{H\sqrt{\theta_{\text{IFOV}}}} \Rightarrow \frac{\theta_{\text{IFOV}}}{2\pi k t_{\text{d}}} \geq \frac{V}{H} \qquad (4.4.5)$$

where w is the width of the scan line. This means that the spatial resolution is increased when t_{d} is decreased if the speed of the detector is increased while maintaining V and H.

4.5 Optical Detectors

The detector location in the optic sensor chain follows the collimation, filtration and isolation of the light. In this section, a review of semiconductors and the photoelectric effect are conducted in addition to the presentation of some exotic materials such as indium antimonide (InSb) and mercury cadmium telluride (HgCdTe) which are used in IR detection. The classic silicon (Si) is mainly limited to visible detection. These detectors have some critical thermal requirements for operation. Therefore an introduction on thermal control is presented.

4.5.1 Semiconductors

In a pure undoped semiconductor, i.e. an intrinsic semiconductor, electrons from the valence band can be excited across that band gap into the conduction band to leave holes and to produce a current. These electrons and holes can move across the material under the influence of an external voltage. The n-type and p-type semiconductors are materials where extra energy levels are added when impurities or dopants are added to an intrinsic semiconductor. For the n-type semiconductor the extra energy level (donor level) is an electron energy level located near the top of the band gap, where electrons can easily be excited into the conduction band. For the p-type material the extra energy level is a hole energy level located in the bottom of the band gap by the valence band which allows the excitation of valence band electrons, leaving mobile holes in the valence band. The effective Fermi level is shifted to a halfway point between the donor level and the conduction level for the n-type and between the acceptor level and the valence band for the p-type (see Figure 4.38).

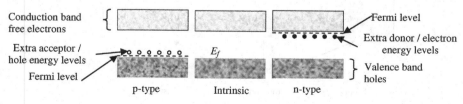

Figure 4.38 Intrinsic, p-type and n-type semiconductors

Table 4.5 Typical photovoltaic detectors, band gaps, cut-off wavelength and operating temperature[5]

Material name	Symbol	E_g(eV)	λ_c (µm)	Operating temp. (K)
Silicon	Si	1.12	1.1	300
Mercury cadmium telluride	HgCdTe	1.00–0.09	1.24–14	20–80
Indium antimonide	InSb	0.23	5.4	30-77
Arsenic doped silicon	Si:As	0.05	24	4
Germanium	Ge	0.88	1.4	300
Lead sulphide	PbS	0.49	2.5	77
Indium arsenide	InAs	0.36	3.3	77
Indium gallium arsenide	InGaAs	0.73	1.7	297

The forbidden energy gap, where no electrons with such energy can exist, is the energy level between the valence band, which is the lowest energy band, and the conduction band. The conduction band is filled when atoms receive external energy such as thermal energy or light. The forbidden gap is different from one material to another. For instance, as presented in Table 4.5, insulators have a large forbidden gap, semiconductors have a small forbidden gap, and metals have no forbidden gap. Figure 4.39 illustrates typical energy levels in metal, semiconductors and insulators.

The Fermi level, E_f, is the highest filled energy level at absolute zero where all electrons fill up the lowest energy levels. As the temperature goes up, electrons are excited to levels above the Fermi

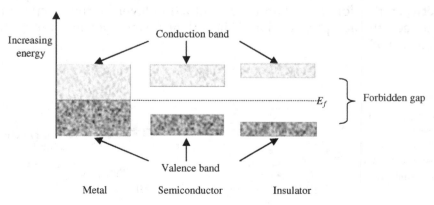

Figure 4.39 Energy levels in matter

level, which change the filling distribution of energy levels. A few levels below the Fermi level are partially empty and those above are partially filled. For semiconductors, the forbidden gap is comparatively narrow. Electrons can be promoted from the valence band to the conduction band across the forbidden energy gap leaving holes instead.

4.5.2 Photoelectric Effect

For an electron to be excited from the conduction band to the valence band the photon energy should be higher than the energy gap of the material:

$$E_p = hf \geq E_g \Rightarrow \lambda_c = h \cdot c / E_g = 1.238 / E_g \qquad (4.5.1)$$

where λ_c is the cut-off wavelength in micrometers, h is the Planck constant, f is the frequency and E_g is the band gap energy of material in electronvolts. The cut-off wavelength is the longest wavelength at which the detector can detect optical radiation.

The photoelectric effect is the phenomenon that occurs when light with a high enough energy hits a metal plate or other reflective material and electrons are ejected. The photon energy is then

$$E_p = hf = W + K_e \qquad (4.5.2)$$

where W is the work function, which is the energy necessary to eject the electron and K_e is the kinetic energy that results from the excess photon energy. This K_e in electrical terms will be equal to the electron's charge $e = 1.610^{-19}$C multiplied by the stopping potential V_{stop}

$$hf = W + V \cdot e \qquad (4.5.3)$$

where f is the frequency of the light. When $V_{stop} = 0$, $f = f_{cut-off}$, and $hf_{cut-off} = W$. In a photovoltaic device, electrons are collected in a circuit and form a current. Thermally generated electrons are indistinguishable from photo-generated electrons. They constitute a noise source known as 'dark current' and it is important that detectors are kept cold to reduce this noise. 1.26 eV corresponds to the energy of light with a wavelength of 1 mm. Beyond this wavelength silicon becomes transparent and CCDs constructed from silicon become insensitive. Most metals have a work function on the order of several

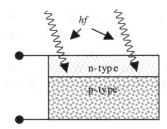

Figure 4.40 n–p junction

electron volts. Any frequency lower than the cut-off value or any wavelength greater than the cut-off value will not eject electrons from the metal.

Photodiodes, also called photovoltaic detectors, are photon-sensitive diodes where current or voltage is produced when optical radiation is absorbed at the n – p junction as illustrated in Figure 4.40.

$$i_g = \eta \phi q$$
$$\eta = n_e/n_p$$

(4.5.4)

where i_g is the photo-generated current produced by photons, η is the quantum efficiency (the efficiency of converting photons to electrons), ϕ is the photon flux, n_p is the number of photons in the active area and n_e is the number of electrons generated.[5] Incident photons on a material must have enough energy to produce electrons.

4.5.3 Performance Criteria for Detectors

There are many performance parameters that describe a detector, such as the spectral response, response time, the sensitivity, the noise figure, quantum efficiency, the noise equivalent power (NEP), the noise equivalent input (NEI), and responsivity. The spectral response of a sensor is the portion of the electromagnetic spectrum where the detector is sensitive. Quantum efficiency is the number of photoelectrons emitted per incident photon expressed as a percentage. A quantum efficiency of 100% means that one electron is emitted for each incident photon. NEP is a measure of the minimum power that can be detected. It is a function of the power incident on the detector that would produce a signal to noise ratio of one, so

$$NEP = V_{noise}/R$$
$$D = 1/NEP \qquad (4.5.5)$$

where R is the responsivity (V/W), and D is the detectivity. The smaller the NEP, the better the sensor. Therefore the higher D the better the detection or sensing. Detector sensitivity is gauged by the figure of merit. The NEI is the radiant power per unit area of the detector required to produce an SNR of one:

$$NEI = NEP/A_d \qquad (4.5.6)$$

where A_d is the detection area. The quantum efficiency of the photocathode is also a figure of merit for photoemissive devices. Photon detectors are sensitive to photons with energies greater than a given intrinsic or gap energy of the detector materials. The sensitivity often increases linearly with increasing wavelength. The responsivity is a quantification of the amount of output seen per watt of radiant optical power input. It is the ratio of the detected signal output to the radiant power input. For photoconductive and photovoltaic detectors the responsivity is measured in RMS volts of the signal V_{sig} per RMS watts of the incident power[5]:

$$R = V_{sig}/W(\lambda)A_d \qquad (4.5.7)$$

The detector area is determined by

$$A_{det} = x^2 = (\beta f)^2 = \beta^2 f^2 = \alpha_d f^2 \qquad (4.5.8)$$

where $\alpha_d = \beta^2$ is the field of view in steradians, f is the focal length of the lens and β is the angle of view subtended by the detector of side x.

4.5.4 Detector Readout

In IR imaging, each pixel has its own transistor for signal amplification and readout. The combination of processing elements and photovoltaic elements allows real time partial readouts and processing. A specific pixel could be isolated for readout without impacting the adjacent ones. This is different from the classic CCD where charges in an array of pixels are amplified at the edge of the pixel set. The photodiode generates a signal charge proportional to the amount of light it receives through photoelectric conversion. This signal is then amplified using a metal oxide semiconductor field effect transistor (MOSFET) amplifier or a junction field effect transistor (JFET) amplifier.

4.5.5 InSb Photodiode

Indium antimonide is widely used for the TIR portion of the spectrum–from 3 to 5 μm wavelength range. The p-type (n-type) material is formed if more indium (antimony) is used. Therefore a p–n junction is formed by adjusting the InSb ratio. For 77 K, liquid nitrogen temperature, 0.23 eV gives $\lambda_c = 5.5$ μm.

4.5.6 HgCdTe Photodiode

Mercury cadmium telluride (HgCdTe) detectors are sensitive up to 13μm and have a wide energy gap semiconductor CdTe with a semi-metallic compound HgTe, which is considered a semiconductor with a negative energy. A negative energy gap means that the conduction band is lower than the top of the valence band. The energy gap is varied by changing the HgTe and CdTe proportions. $Hg_{1-x}Cd_xTe$ varies from -0.3 eV to 1.605 eV depending on x, the mole fraction ratio of Cd to Hg. As x decreases the cut-off wavelength increases which is very convenient for IR detection. An approximation[5] of the energy gap as a function of x and T is

$$E_g = -0.302 + 1.93x + 5.35.10^{-4}T(1-2x) - 0.310x^2 + 0.832x^3$$
$$\lambda_c = 1.24/E_g. \hspace{2cm} (4.5.9)$$

In the fabrication process, homogeneity in the detector is a challenge. Therefore, detector arrays are small and each different detector may have a distinct cut-off frequency. Figure 4.41 shows that for higher λ_c lower temperatures are needed. A lower operating temperature would also reduce thermal noise.

By modifying the ratio of mercury and cadmium the band gap energy is adjusted. The wavelength λ_c is determined by x, which determines the operating temperature (Table 4.6).

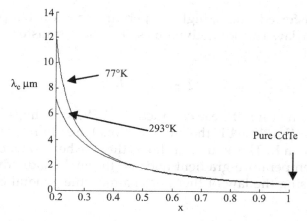

Figure 4.41   Effect of x and T on λ_c the cut-off wavelength[†]

Table 4.6   Demonstration of tunable Mer-Cad-Tel bandgap

x	0.196	0.21	0.295	0.395	0.55	0.7
E_g (eV)	0.09	0.12	0.25	0.41	0.73	1
λ_c μm	14	10	5	3	1.7	1.24

4.5.7 Thermal Control

There are many sources for heat accumulation on a spacecraft such as direct heat from the sun or indirect heat via reflection, onboard operation, power source or payload electronics. In order to ensure the proper function of subsystems, to minimize noise and to increase reliability, thermal control systems (TCS) have to be properly design-ed and implemented. The goal of the TCS is to maintain a specific thermal environment at the subsystem level to fulfill operating temperature requirements in all modes: operating, standby or non-operating. For instance, for IR detectors, the operating temperature of HgCdTe or InSb is in the 70 K range.

In spacebased sensors, there are mainly two types of heat trans-fer: conduction and thermal radiation. Conduction is heat transfer by means of molecular agitation within a material without any motion of the material. It takes place in solids, liquids and gases.

[†] Figure 4.41 was generated using the following Matlab code. $T = 293; x = 0 : 0.01 : 1; E_g = -0.302 + 1.93^*x + (5.35e-4)^*T^*(1 - 2^*x) - 0.310^*x.^*x + 0.832.^*x.^*x.^*x;$ lamdac$= 1.238./E_g$; hold on; plot(x, lamdac, $'k'$); axis([0. 2 1 0 14]).

Heat is transferred from a high temperature to a low temperature as per the first law of thermodynamics. The heat transfer rate in one dimension is

$$Q = -\kappa A \partial_x T \tag{4.5.10}$$

where Q is the rate of heat conduction in W, κ is the thermal conductivity in $Wm^{-1}K$, A is the cross-sectional area in m^2, and T is the temperature in K. This equation shows that the heat transfer depends upon the temperature gradient and the thermal conductivity of the material. Using the law of thermodynamics, the general conduction equation is

$$\nabla^2 T = \frac{1}{\alpha_d} \frac{\partial T}{\partial t} - \frac{g_{int}}{\kappa}$$
$$\alpha_d = \kappa / \rho c_p \tag{4.5.11}$$

where ρ is the density in kgm^{-3}, c_p is the specific heat in $Jkg^{-1}k^{-1}$, g_{int} is the internal volume source in Wm^{-3} and α_d is the thermal diffusivity. Material properties include α_d and c_p (see Table 4.7) and using the appropriate material for heat conduction is crucial in thermal design. This equation shows that by appropriately selecting α_d and c_p thermal conductivity can be optimized.

For thermal radiation transfer, objects at temperature T emit or absorb EM radiation (section 4.1.11). The emitted heat Q_E and the absorbed heat Q_A by EM radiation are

$$Q_E = \varepsilon A \sigma \left(T^4 - T_{env}^4 \right)$$
$$Q_A = \alpha A S \tag{4.5.12}$$

where ε is the emissivity which is the efficiency of a surface to radiate heat, A is area of the object, σ is Boltzmann's constant, α is the absorptivity, T_{env} is the temperature of the environment and S is the flux such as solar flux.

Table 4.7 Specific heat capacities for different materials

Material	Copper	Aluminum	Beryllium	Titanium	Tungsten	Hydrazine	GaAs	Silicon	Kapton
Specific heat, c_p	390	920	1800	520	142	3100	335	712	1006

Thermal control systems (TCS) are achieved either by using a passive thermal control or active thermal control (ATC). For passive thermal control the absorptivity, α, and the emissivity, ε, are to be optimized to achieve a thermal design range. This could be done by properly choosing the paint and coating patterns. This could also be done by adding a thermal blanket such as multi-layer insulation (MLI) which is aluminized Mylar or kapton layers separated by a thin net of material such as nylon. Orientation of the spacecraft to face the Sun or to face away from the Sun also controls temperature. ATC is used when passive methods are insufficient. There are many approaches for ATC such as heaters, heat pipes, radiators (louvers or venetian blinds), thermoelectric cooling (good for spot cooling, such as IR detectors), cryogenic cooling and venting.

Heat pipes are based on heat dissipation by evaporation and condensation. The liquid in the pipe absorbs the thermal energy and is transformed to a gas which is transferred to the other end of the pipe. At that end the gas condenses and cools to the liquid state by releasing its heat to a radiator. This liquid then returns to the hot area. With a heat pipe, heat dissipation of a spacecraft is equal to

$$Q_d = Q_{hp} + Q_p + Q_r = c_{hp}(T_e - T_c) + \varepsilon_p A_p \sigma T^4 + \varepsilon_r A_r \sigma \eta T^4 \quad (4.5.13)$$

where Q_d is dissipated heat, Q_{hp} is radiated heat using heat pipes, Q_p is radiated using paint, Q_r uses a passive radiator, c_{hp} is the heat pipe thermal conductance, T_e and T_c are evaporator surface temperature and condenser temperature, ε_p and ε_r are the emissivities of the paint and the radiator, A is the area, η is the efficiency and σ is the Stefan–Boltzmann constant.

4.6 Landsat 7: ETM+

The enhanced thematic mapper plus (ETM+) sensor, is part of the Landsat 7 spacecraft (see Figure 4.42). It was built based on the heritage from the thematic mapper (TM) built for Landsats 4 and 5 and the enhanced thematic mapper (ETM) lost during the Landsat 6 failure. The altitude of the spacecraft is 705 km in a near polar, sun-synchronous and circular orbit with an orbit inclination of 98.2 degrees. The ETM+ (see Figure 4.43) is an eight-band multispectral

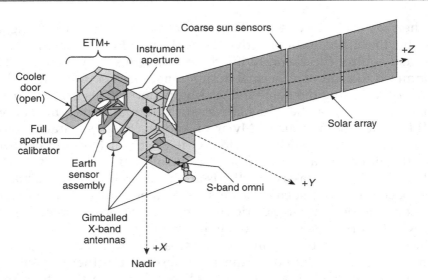

Figure 4.42 Landsat 7 and the ETM+ sensor (Courtesy of Landsat Project Science Office)

scanning radiometer.[16] It provides high resolution image information of the Earth's surface. Cross-track scanning of the ETM+ allows a 185 km swath. Along-track is provided by the orbital motion of the spacecraft. Table 4.8 summarizes some of the ETM+ characteristics.

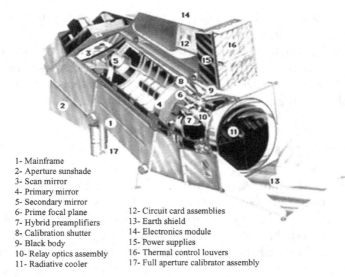

1- Mainframe
2- Aperture sunshade
3- Scan mirror
4- Primary mirror
5- Secondary mirror
6- Prime focal plane
7- Hybrid preamplifiers
8- Calibration shutter
9- Black body
10- Relay optics assembly
11- Radiative cooler
12- Circuit card assemblies
13- Earth shield
14- Electronics module
15- Power supplies
16- Thermal control louvers
17- Full aperture calibrator assembly

Figure 4.43 ETM+ sensor (Courtesy of Landsat Project Science Office)

Table 4.8 Radiometric characteristics of the ETM+[16]

Band #	Spectral range (microns)	EM region	Detector type	Pixel size	# detector element	Generalized application details
1	0.45–0.52	Visible blue	Si	30 × 30 m	16	Coastal water mapping, differentiation of vegetation from soils
2	0.52–0.60	Visible green	Si	30 × 30 m	16	Assessment of vegetation vigor
3	0.63–0.69	Visible red	Si	30 × 30 m	16	Chlorophyll absorbtion for vegetation differentiation
4	0.76–0.90	NIR	Si	30 × 30 m	16	Biomass surveys and delineation of water bodies
5	1.55–1.75	MIR	InSb	30 × 30 m	16	Vegetation and soil moisture measurements; differentiation between snow and cloud
6	10.40–12.50	TIR	HgCT	60 × 60 m	8	Thermal mapping, soil moisture studies and plant heat stress measurement
7	2.08–2.35	MIR	InSb	30 × 30 m	16	Hydrothermal mapping
8	0.52–0.90	Visible and NIR	Si	18 × 18 m	32	Large area mapping, urban change studies

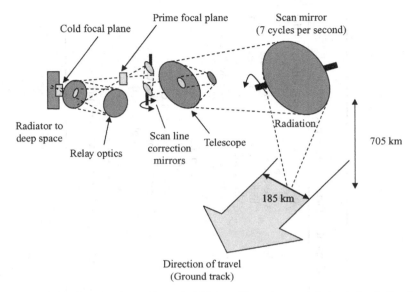

Figure 4.44 Optical path of ETM+ (Courtesy of Landsat Project Science Office)

Figure 4.44 shows the optical path of ETM+. The scanning mirror scans the ground in the across-track direction. Part of the incoming energy is focused onto a prime focal plane via a scan line corrector using a Ritchey–Chretien telescope. In this plane, bands 1 through 4 and band number 8 are monitored using a monolithic silicon detector. The other part of the incoming energy is optically redirected to the cold focal plane where the rest of the detectors are located. Radiative coolers are used to maintain a 91 K focal plane temperature as required for InSb and HgCT detectors.

4.7 ASTER

ASTER stands for advanced spacebased thermal emission and reflection radiometer. ASTER is an imaging instrument that is part of the Terra spacecraft (see Figure 4.45). Terra was launched in December 1999. It is used to obtain detailed maps of land surface temperature, emissivity, reflectance and elevation to monitor

Figure 4.45 Terra spacecraft (Courtesy of NASA/GSFC/METI/ERSDAC/ JAROS, and U.S./Japan ASTER Science Team)

climatological processes, soil moisture, ocean currents, vegetation, volcanos and thermal pollution.

Aster is an ensemble of three instruments: TIR, SWIR and VNIR.[17] The following sections will present some characteristics of each of these three instruments. Each of these instruments is dedicated to a specific spectral range. Table 4.9 shows the various bands of each instrument as well as the appropriate detector used.

4.7.1 ASTER: TIR

Thermal IR (TIR) instruments measure emissive energy or the temperature of a target in the 8 to 14 μm range. Figure 4.18 shows that these two bands are within the atmospheric windows. Using TIR, surface spectral emittance and temperature distribution on a surface can be monitored.

Figure 4.46 shows some of the components of the ASTER TIR instrument. It includes a telescope, cryocooler, scan mirror and a reference plate. The telescope is a fixed Newtonian catadioptric system. A whiskbroom scanning mirror is used for calibration and pointing. In the calibration mode, the mirror alternates between

Table 4.9 VNIR, visible and near infrared; SWIR, shortwave infrared; TIR, thermal infrared[17]

Characteristic	VNIR	SWIR	TIR
Spectral range	Band 1: 0.52 - 0.60 μm	Band 4: 1.600 - 1.700 μm	Band 10: 8.125 - 8.475 μm
	Band 2: 0.63 - 0.69 μm	Band 5: 2.145 - 2.185 μm	Band 11: 8.475 - 8.825 μm
	Band 3: 0.76 - 0.86 μm	Band 6: 2.185 - 2.225 μm	Band 12: 8.925 - 9.275 μm
	Band 3: 0.76 - 0.86 μm	Band 7: 2.235 - 2.285 μm	Band 13: 10.25 - 10.95 μm
		Band 8: 2.295 - 2.365 μm	Band 14: 10.95 - 11.65 μm
		Band 9: 2.360 - 2.430 μm	
Ground resolution	15 m	30 m	90 m
Cross-track pointing (deg.)	±24	±8.55	±8.55
Cross-track pointing (km)	±318	±116	±116
Swath width (km)	60	60	60
Detector type	Si	PtSi – Si	HgCdTe
Quantization (bits)	8	8	12

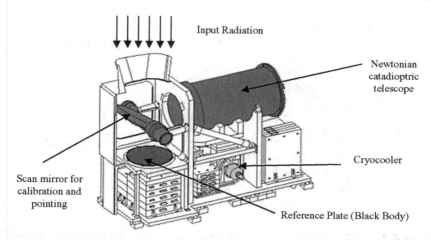

Figure 4.46 ASTER TIR instrument (Courtesy of NASA/GSFC/METI/ ERSDAC/JAROS, and U.S./Japan ASTER Science Team)

nadir pointing and reference plate direction. This allows the system to conduct measurements on a known controlled reference plate (see section 4.1.11). The reference plate is the equivalent of a black body. It has a high emissivity and is thermally controlled. It is used to calibrate the instrument between measurements or observations. This allows an assessment of any instrument noise, offset calibration, drift and gain. In the pointing mode, the mirror is nadir pointed with a 17 degree sweep. This allows a wide cross-track scanning. The TIR instrument has 50 HgCdTe detectors, ten detectors are in a staggered array for each of the five bands (see Table 4.9). Optical band-pass filters are used over each detector element. The cryocooler is used to cool the HgCdTe detectors (as per Table 4.5) and requires low temperature for operation. The cryocooler is a mechanical split Stirling cycle cooler.

4.7.2 ASTER: SWIR

SWIR stands for short wavelength IR sensor. This instrument operates in six spectral bands in the near IR (see Table 4.9). Figure 4.47 shows the various components of the SWIR. The telescope is a single fixed aspheric refractor, nadir-pointing that provides 30 m of resolution. SWIR's detectors are cooled to 80 K using a mechanical split Stirling cycle cooler. A scanning mirror is used for pointing within 19° of FOV. It is also used to steer periodically between the

Scanning mirror

Telescope

Cryocooler

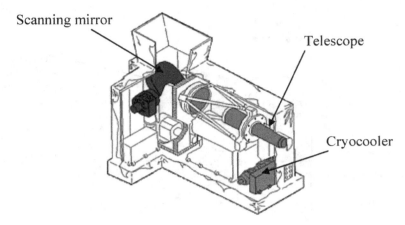

Figure 4.47 SWIR instrument (Courtesy of NASA/GSFC/METI/ERSDAC/ JAROS, and U.S./Japan ASTER Science Team)

calibration source and the target to the telescope. The calibration source is composed of two halogen lamps.

4.7.3 ASTER: VNIR

VNIR stands for visible near IR sensor. VNIR measures reflected EM waves in the three spectral bands at visible and near IR wavelengths, with a resolution of 15 m. Nevertheless, when measuring hot targets such as fires, measurements are conducted using the emitted energy from such a target. Figure 4.48 shows the VNIR instrument. VNIR

Nadir Looking Telescope

Backward Looking Telescope.

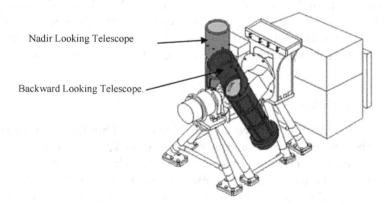

Figure 4.48 VNIR instrument (Courtesy of NASA/GSFC/METI/ERSDAC/ JAROS, and U.S./Japan ASTER Science Team)

has the capability of stereo observation of band 3. This is done using two telescopes. One is pointing in the nadir direction and the other one is the backward-looking telescope which provides a second view of the target at band 3. Further cross-tracking capability can be done by allowing a rotation of 48 degrees of the ensemble of the two telescopes.

The nadir-looking telescope is is a reflecting–refracting improved Schmidt design. Three 5000 silicon charge coupled detector line arrays are placed in the focal plane of this telescope. It is periodically calibrated by two on-board halogen lamps that are located in the optical path. A combination of dichroic elements and interference filters allows the separation of the three bands. The backward-looking telescope, which is used only for stereo observation, is of the same design as the nadir telescope and contains only a single silicon charge coupled detector line array and no calibration lamps.

4.8 GOES

The geostationary environmental operational satellite (GOES, see Figure 4.49), is designed to monitor the Earth's surface, the space environment and weather patterns. GOES views the US at an altitude of 35 790 km above the equator. GOES sounder and GOES imager are two of the payloads of interest that GOES carried.[15] Refer to Figure 4.49 to see where they are located in reference to the spacecraft. The following sections will introduce these sensors.

4.8.1 GOES-I Imager

Figure 4.50 illustrates the various subsystems of the GOES imager and Table 4.10 covers the GOES imager characteristics. The telescope is a Cassegrain telescope (refer to section 1.6). It is used in conjunction with two-axis gimbaled scanning mirrors. This instrument can scan 3000 × 3000 km center over the US in less than 1 minute. The telescope, scan mirror assembly and detectors are mounted on a base plate outside the main structure of the spacecraft, as illustrated in Figure 4.49. Thermal control is accomplished using shields and louvers.

The detector layout is illustrated in Figure 4.51. The size of each individual detector for each band is represented by a square.

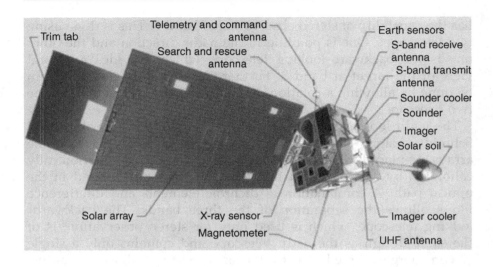

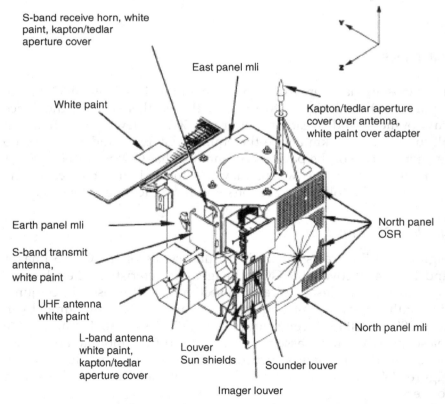

Figure 4.49 GOES spacecraft (Courtesy of NASA-Goddard Space Flight Center)

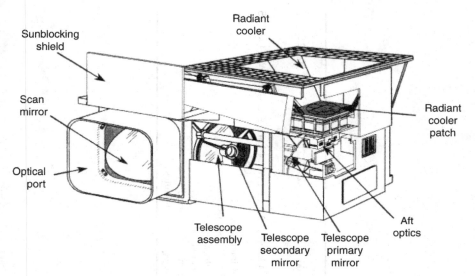

Figure 4.50   GOES-I imager (Courtesy of NASA-Goddard Space Flight Center)

4.8.2 GOES-I/M Sounder

The GOES Sounder is a 19-channel, discrete-filter radiometer covering the spectral range from the visible channel wavelengths to 15 microns (refer to Table 4.11). It is designed to provide data from which atmospheric temperature, moisture profiles, surface and cloud-top temperatures, and ozone distribution can be deduced by mathematical analysis. It operates independently of, and simultaneously with, the imager using a similarly flexible scan system. The Sounder's multi-element detector array assemblies simultaneously sample four separate fields or atmospheric columns. A rotating filter wheel, which brings spectral filters into the optical path of the detector array, provides the IR channel definition (see Figure 4.52).

4.9 DSP and SBIRS

The Defense Support Program (DSP) is a spacebased system that detects, reports and tracks missile and space launches in real time. This is done using state-of-the art IR sensors. IR sensors are suitable

Table 4.10 GOES imager characteristics[15]

Channel number	1	2	3	4	5
Wavelength range (um)	0.55–0.75	3.80–4.00	6.50–7.00	10.20–11.20	11.50–12.50
Instantaneous geographic field of view (IGFOV) at nadir	0.57 km EW × 1 km NS	2.3 km EW × 4 km NS	2.3 km EW × 4 km NS	2.3 km EW × 4 km NS	2.3 km EW × 4 km NS
Detector type	Silicon	InSb	HgCdTe	HgCdTe	HgCdTe
Radiometric calibration	Space and 290 K infrared internal back body				
Calibration frequency	Space: 2.2 sec (full disc), 9.2 or 36.6 sec (sector/area) Infrared: 30 minutes typical				
System absolute accuracy	IR channels: less than or equal to 1 K Visible channel: 5% of maximum scene irradiance				
Imaging rate	Full Earth disc, less than or equal to 26 minutes				

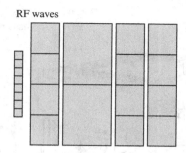

RF waves

Figure 4.51 Detector layout

for such applications because of the detection approach that the DSP is using. Indeed, the signature of missile or space launches is identified by the byproduct of the propulsion system located in the missile plume. The plume composition depends on the fuel used in the propulsion system of these missiles. The plume extent depends on the combustion process and on the altitude. DSP also has the capability of detecting nuclear detonations.

Figure 4.53 shows the satellite vehicle and some of its subsystems. There are many DSP spacecraft that are currently on orbit. At least

Table 4.11 GOES Sounder characteristics[15]

Channel	GOES Sounder spectral peak	Detector type	Band	Purpose
1	14.71	HgCdTe	Carbon dioxide	Stratospheric temperature
2	14.37		Carbon dioxide	Tropospheric temperature
3	14.06		Carbon dioxide	Upper-level temperature
4	13.96		Carbon dioxide	Mid-level temperature
5	13.37		Carbon dioxide	Mid-level temperature
6	12.66		Water Vapor	Total precipitable water
7	12.02		Window	Surface temperature, moisture
8	11.08		Window	Surface temperature
9	9.71		Ozone	Total ozone
10	7.43		Water vapor	Low-level temperature
11	7.02		Water vapor	Mid-level temperature
12	6.51		Water vapor	Upper-level temperature
13	4.57	InSb	Carbon dioxide	Low-level temperature
14	4.52		Carbon dioxide	Mid-level temperature
15	4.45		Carbon dioxide	Upper-level temperature
16	4.13		Nitrogen	Boundary layer temperature
17	3.98		Window	Surface temperature
18	3.74		Window	Surface temperature, moisture
19	0.969	Si	Visible	Clouds

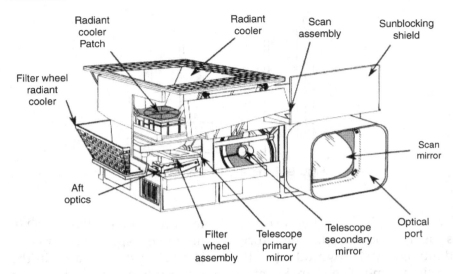

Figure 4.52 GOES-I Sounder (Courtesy of NASA-Goddard Space Flight Center)

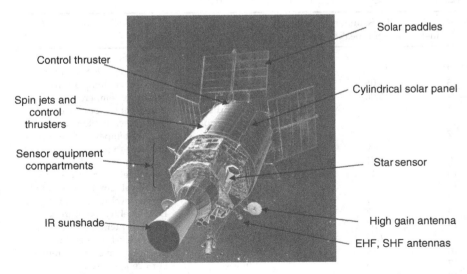

Figure 4.53 DSP spacecraft (Courtesy of Northrop Grumman Corporation, USA ©)

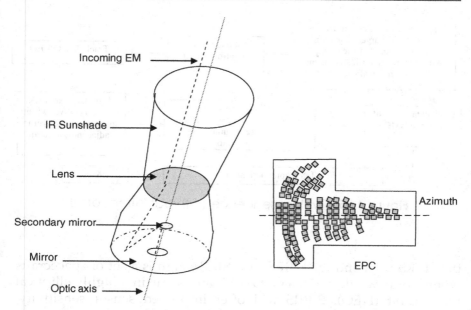

Figure 4.54 Typical EPC and DSP telescope configuration

three of them are needed to ensure global coverage. Satellites are placed in a circular, equatorial, 35 000 km altitude geosynchronous orbit. This orbit allows a continuous monitoring of the same area. The IR telescope is pointed toward the Earth and rotates about its earth-pointing axis at six revolutions per minute. To provide a scanning motion for the IR sensor. This spin provides the tilted boresight on the DSP.

A DSP-1 photo-electric PEC array contains more than 6000 detector cells, as illustrated in Figure 4.54. The cells are sensitive to IR energy in the IR spectrum. As the satellite rotates, the Earth's surface is scanned (see Figure 4.54). With the PEC array scanning the FOV, a cell passing across an IR source will develop a signal with an amplitude that is proportional to its intensity. The signal is then amplified and relayed to one of the DSP ground stations.

The spacebased IR system (SBIRS) would be a system that replaces the DSP system. Even though the status of this program is currently uncertain, a brief description of one of the conceptual designs is presented in Figure 4.56. SBIRS would be composed of two systems,

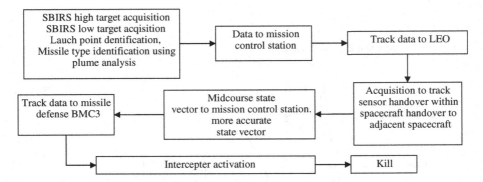

Figure 4.55 Typical sequence of threat intercepted

the SBIRS high and the SBIRS low. SBIRS high is a set of spacecrafts where some would be in a GEO orbit and some in a highly elliptical Earth orbit (HEO). SBIRS will offer improved sensor sensitivity, mid-course tracking and discrimination, higher revisit rate, a more accurate state vector handover to the intercept, faster missile reporting, multi-target tracking, a more accurate impact point and a more accurate launch point. This will lead to an improvement in the launch point and impact point, and the ability to detect shorter range missiles. Figures 4.55 and 4.56 illustrate a typical threat that is detected, identified, reported, tracked and intercepted.

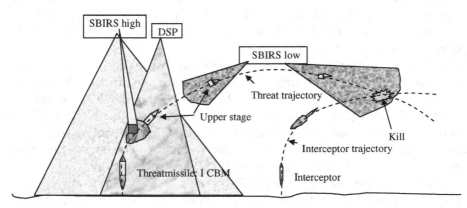

Figure 4.56 SBIRS: an example of coordination in a typical threat situation[20]

Bibliography

1. Asrar, G. (1989) *Theory and Applications of Optical Remote Sensing*, John Wiley and Sons.
2. Barton, D. K. (1988) *Modern Radar System Analysis*, Artech House, Inc.
3. Campbell, J. B. (1996) *Introduction to Remote Sensing*, 2nd edn, Taylor and Francis, London.
4. Colwell, R. N. (ed) (1983) *The manual of remote sensing*, American Society of Photogrammetry.
5. Dereniak, E. L. and Boreman, G. D. (1996) *Infrared Detectors and Systems*, John Wiley and Sons, Inc.
6. Elachi, C. (1987) *Introduction to the Physics and Techniques of Remote Sensing*, John Wiley and Sons, Inc.
7. Giancoli, D. C. (1998) *Physics*, Prentice-Hall.
8. Hecht, E. (1998) *Optics*, Addison Wesley Longman, Inc.
9. Jelalian, A. V. (1992) *Laser Radar Systems*, Artech House.
10. Karp, S. *et al.* (1988) *Optical Channels*, Plenum Press.
11. Kieffer, R. W. and Lillesand, T. M. (1987) *Remote Sensing and Image Interpretation*, 2nd edn, John Wiley & Sons.
12. Kingston, R. H. (1995) *Optical Sources, Detectors and Systems, Fundamentals and Applications*, Academic Press, Inc.
13. Larson, W. J. and Wertz, J. R. (1992) *Space Mission Analysis and Design*, Microcosm, Inc. and Kluwer Academic Publisher.
14. Lillesand,T. M. and Kiefer, R. W. (2000) *Remote Sensing and Image Interpretation*, 4th edn, John Wiley and Sons, Inc.
15. NASA GSFC (1996) *GOES I-M DataBook*.
16. NASA GSFC (1999) *Landsat 7 Science Data Users Handbook*.
17. NASA JPL (2002) *Aster User Handbook*.
18. Pease, C. B. (1991) *Satellite Imaging Instruments*, Ellis Horwood Limited.
19. Rees, W. G. (1990) *Physical Principles of Remote Sensing*, Cambridge University Press.
20. Simmons, F. S. (2000) *Rocket Exhaust Plume Phenomenology*, The Aerospace Corporation.
21. Ulaby, F. T., Moore, R. K., and Fung, A. K. (1981) *Microwave Remote Sensing: Active an Passive*, Volume I, Addision-Wesley Publishing Company, Inc.
22. Ulaby, F. T., Moore, R. K., and Fung, A. K. (1986) *Microwave Remote Sensing: Active an Passive*, Volume III, Addison-Wesley Publishing Company, Inc.

5

Passive Microwave Sensors

In the microwave domain the atmosphere is transparent to microwaves under virtually all climatic conditions (see Figure 4.17). Microwave sensors operate in the wavelength range from 1 mm to 1 m which corresponds to a frequency range of 300 GHz to 0.3 GHz. Attenuation in this window is very low. For wavelengths smaller than 1 mm, the atmosphere between the spacecraft and the target is essentially opaque because H_2O and other atmospheric particles absorb all of the energy. In the microwave band antennas are used, instead of semiconductor materials, as transducers that transfer the EM waves from/into the free space. To guide the EM signal along a predetermined path, waveguides are used.

The general categories of microwave sensors encompass both passive and active sensors. Chapter 6 will focus on active microwave sensors, such as radars in the form of altimeters and SAR. In this chapter, passive microwave sensors such as radiometers will be presented. This will be preceded by an overview on antennas, waveguides and phased arrays.

5.1 Antenna

An antenna can be a dual function component that can receive or transmit EM waves. In the transmitting mode, the antenna transforms current or electrical energy into a radiated EM wave. The

Spacecraft Sensors.   Mohamed M. Abid
© 2005 John Wiley & Sons, Ltd

reverse process occurs when the antenna is in the receiving mode. Antennas are also considered to be impedance matching devices. During transmission, radiated energy is concentrated into a well-defined beam. Depending on the geometric orientation of the antenna, it points energy in the desired direction to illuminate a selected target. The transmission medium could be a line or wave-guide and the transmitting/receiving antenna could be a wire, a microstrip patch or a horn. Some of these will be explored in the following section.

5.1.1 Vector Potential

Using the following relations

$$\frac{\partial}{\partial t} = j\omega$$
$$\mu H = B \qquad\qquad (5.1.1)$$
$$\varepsilon E = D$$

Maxwell's equations become

$$\nabla \times E = -j\omega\mu H$$
$$\nabla \times H = j\omega\varepsilon E + J \qquad\qquad (5.1.2)$$
$$\nabla \bullet D = \rho; \nabla \bullet H = 0,$$

where E is the electric vector, D is the electric displacement, B is the induction vector, H is the magnetic field strength vector, ρ is the electric charge density, J is the current density vector, $\nabla\bullet$ is divergence, $\nabla\times$ is the curl, ε is the permittivity and μ is the permeability. Using magnetic potential vectors A, electric scalar potential Φ, and the current continuity equation,

$$\nabla \bullet J = -j\omega\rho. \qquad\qquad (5.1.3)$$

The divergence of a vector is scalar. One can derive the following relations:

$$\nabla \bullet H = 0 \Rightarrow H = \frac{1}{\mu}\nabla \times A \qquad\qquad (5.1.4)$$

$$\nabla \times (E + j\omega A) = 0 \Rightarrow E + j\omega A = -\nabla\Phi \Rightarrow E = -j\omega A - \nabla\Phi. \qquad (5.1.5)$$

The gradient of a scalar is a vector. The wave equation for the magnetic vector potential is

$$\nabla^2 A + \omega^2 \mu \varepsilon A - \nabla(j\omega\mu\varepsilon\Phi + \nabla \bullet A) = -\mu J. \tag{5.1.6}$$

Using Lorentz Gauge, which is basically another form of the continuity equation:

$$\nabla \bullet A = -j\omega\mu\varepsilon\Phi \tag{5.1.7}$$

the vector wave equation is

$$\nabla^2 A + \omega^2 \mu \varepsilon A = \nabla^2 A + k^2 A = -\mu J; k = \omega\sqrt{\mu\varepsilon} \tag{5.1.8}$$

The wave equation for the electric scalar potential is

$$\nabla \bullet E = \frac{\rho}{\varepsilon} \Rightarrow \nabla^2 \Phi + j\omega\nabla \bullet A = \nabla^2 \Phi + k^2 \Phi = -\frac{\rho}{\varepsilon}; k = \frac{2\pi}{\lambda} \tag{5.1.9}$$

where k is the wave number. Sometimes the variable β is used instead of k. These wave equations are inhomogeneous Helmholtz equations which apply to regions where currents and charges are not zero. The general solutions are

$$A = \int\int\int_V \mu J(r) \frac{e^{-jkR}}{4\pi R} dv$$
$$\Phi = \int\int\int_V \rho(r) \frac{e^{-jkR}}{4\pi\varepsilon R} dv; R = |r_t - r|. \tag{5.1.10}$$

These integrals are extended to all points over the antenna body where the sources (current density, charge) are not zero. Figure 5.1 shows an antenna body with an electric charge density ρ and a current density J that each antenna volume δv of the antenna would radiate.

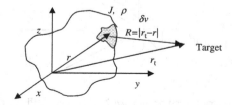

Figure 5.1 Radiating antenna body

The effect of each volume element of the antenna is to radiate a radial wave proportional to

$$\frac{e^{-jk|r_t-r|}}{|r_t - r|} \tag{5.1.11}$$

This quantity is commonly known as the free space Greene's function.

5.1.2 Infinitesimal Antenna

Figure 5.2 illustrates an infinitesimal antenna. The current flowing in the infinitesimal antenna is assumed to be constant and oriented along the z-axis, so

$$\left.\begin{array}{c} I = \Delta s \cdot J(r) \\ \Delta v = \Delta s \cdot \Delta z \end{array}\right\} \Rightarrow \Delta v \cdot J(r) = |I|\Delta z \cdot i_z \tag{5.1.12}$$

where i_z is the unit vector in the z direction. The evaluation of the integral at the origin leads to the solution of the wave equation for the magnetic vector potential.

$$A = \int\int\int_V \mu J(r) \frac{e^{-jkR}}{4\pi R} dv = \mu|I|\Delta z \frac{e^{-jkr_t}}{4\pi r_t} i_z. \tag{5.1.13}$$

Using cylindrical coordinates this equation becomes

$$A = \mu|I|\Delta z \frac{e^{-jkr_t}}{4\pi r_t} i_z = \mu|I|\Delta z \frac{e^{-jkr_t}}{4\pi r_t} (i_r \cos\theta - i_\theta \sin\theta). \tag{5.1.14}$$

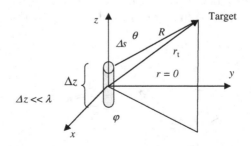

Figure 5.2 Infinitesimal antenna with the elevation angle θ and the azimuth angle φ

Using equation (5.1.2), H and E become

$$H = \frac{1}{\mu} \nabla \times A \Rightarrow H_\varphi = -k^2 |I| \Delta z \frac{e^{-jkr_t}}{4\pi} \left(\frac{1}{jkr_t} + \frac{1}{(jkr_t)^2} \right) \sin\theta \qquad (5.1.15)$$

$$\Rightarrow \begin{cases} E_r = -\eta k^2 |I| \Delta z \frac{e^{-jkr_t}}{4\pi} 2 \left(0 + \frac{1}{(jkr_t)^2} + \frac{1}{(jkr_t)^3} \right) \cos\theta \\[4mm] E_\theta = -\eta |I| k^2 \Delta z \frac{e^{-jkr_t}}{4\pi} \left(\frac{1}{jkr_t} + \frac{1}{(jkr_t)^2} + \frac{1}{(jkr_t)^3} \right) \sin\theta \end{cases} ;$$

$$\eta = \sqrt{\frac{\mu}{\varepsilon}} \qquad (5.1.16)$$

where η is the intrinsic impedance of the medium. In free space, $\eta = 377\,\Omega$. The rest of the components H_r, H_θ, and E_φ are null. The term $1/jkr$ is the radiation field. The term $(1/jkr)^2$ is the induction field. The term $(1/jkr)^3$ is the electrostatic field. Depending on the location of the measurement, one field will be dominant. For the near case when $d \ll r_t \ll \lambda$, the $1/r_t^3$ is dominant and therefore the dipole will be seen as an electrostatic source. For large r_t when $d \ll r_t \sim \lambda$, this is the Fresnel case where the field will be seen as an induction field since mainly the $1/r^2$ term is dominant. For $r > \lambda \gg d$, this is the Fraunhofer case where the field will be seen as a radiation field. In both cases d is the antenna size.[6] For a large distance between the antenna and the target and where the antenna could be just considered as a point source, equation (5.1.15) becomes

$$r \gg \lambda \Rightarrow E_\theta \approx \eta jk |I| \Delta z \frac{e^{-jkr_t}}{4\pi r_t} \sin\theta$$

$$H_\varphi \approx jk |I| \Delta z \frac{e^{-jkr_t}}{4\pi r_t} \sin\theta. \qquad (5.1.17)$$

The rest of the components' contributions are negligible. This shows that the radiated fields are perpendicular to each other and to the direction of propagation. Also, they are in phase. The surfaces of constant phase are spherical instead of planar and the wave travels in the radial direction. For a uniform wave, the field intensity is inversely proportional to the distance. The field intensities are not constant on a given surface of constant phase. The intensity depends on the sine of the elevation angle. The radiated power density is

$$P = \frac{1}{2}\text{Re}(E \times H^*) = P_r i_r$$

$$P_r = \frac{1}{2}\eta\left(\frac{k|I|\Delta z}{4\pi r_t}\right)^2 \sin^2\theta$$

(5.1.18)

where the H^* is the complex conjugate of H and Re is the real part. The radiated power is proportional to the inverse square of the target location. The total radiated power P_{tot} is obtained by integrating the radiated power density over the sphere:

$$P_{tot} = \int_0^{2\pi}\int_0^{\pi} r^2\sin\theta|S|d\varphi \cdot d\theta = \frac{4\pi}{3}\eta\left(\frac{k|I|\Delta z}{4\pi}\right)^2.$$

(5.1.19)

The total radiated power, which is also the power delivered by the transmission line to the real part of the equivalent impedance, Z, seen at the input of the antenna, is independent of distance:

$$P_{tot} = \frac{1}{2}|I|^2 Z_{equivalent} = \frac{4\pi}{3}\eta\left(\frac{k|I|\Delta z}{4\pi}\right)^2 \Rightarrow Z_{equivalent} = \frac{2\pi}{3}\eta\left(\frac{\Delta z}{\lambda}\right)^2$$

(5.1.20)

where $Z_{equivalent}$ is the equivalent impedance. The equivalent resistance of the antenna is often called the radiation resistance R_r. In free space:

$$R_r = \frac{2\pi}{3}\sqrt{\frac{\mu_0}{\varepsilon_0}}\left(\frac{\Delta z}{\lambda}\right)^2 = 80\pi^2\left(\frac{\Delta z}{\lambda}\right)^2.$$

(5.1.21)

For an emitting or receiving antenna, the energy that reaches the antenna has two parts: one part that is radiated out and one part that is dissipated as heat. Therefore a radiation resistance, R_R, and loss resistance, R_L, are used to quantify each part. They are defined as

$$P_0 = \frac{1}{2}I_0^2 R_R$$

$$P_L = \frac{1}{2}I_0^2 R_L$$

(5.1.22)

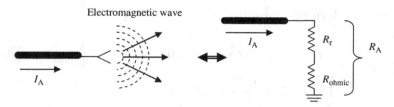

Figure 5.3 Antenna equivalent impedance

where P_0 is the average radiated power and P_L is the average dissipated power. In terms of radiation efficiency η_l

$$\eta_l = P_0/(P_0 + P_L) = R_R/(R_R + R_L). \qquad (5.1.23)$$

The equivalent of a transmitting antenna in circuits is an impedance that dissipates the transmitted power (see Figure 5.3).

The total radiated power is also used to define the average power density emitted by the antenna. The average power density corresponds to the radiation of a hypothetical isotropic antenna, which is used as a reference to understand the directive properties of any antenna. The time-average power density is given by

$$P_{\text{ave}} = \frac{P_{\text{tot}}}{4\pi r^2} = \frac{n}{3}\left(\frac{k|I|\Delta z}{4\pi r}\right)^2. \qquad (5.1.24)$$

For dipole antennas, maximum power or gain is transmitted or received in the perpendicular direction of the antenna and there is no power or gain along the antenna.

In the near field where the target is close to the antenna, as opposed to the far field antenna, where the antenna for an observer appears as a point source or receiver, the radiation pattern is rather complex. The antenna is often considered as a collection of small conductors where each conductor acts as an emitter or a receiver. For space applications, the near field is important to measure and to control because it could have an impact on a payload that is susceptible to electromagnetic interference in the specific operational frequency of the antenna. For the case of an infinitesimal dipole, the near field is derived as follows:

$$kr \ll 1 \Rightarrow \begin{cases} E \approx -j\eta|I|\Delta z \dfrac{e^{-jkr_t}}{4k\pi r_t^3} \sin(\theta)i_\theta - j\eta|I|\Delta z \dfrac{e^{-jkr_t}}{2k\pi r_t^3} \cos(\theta)i_r \\[4mm] H \approx |I|\Delta z \dfrac{e^{-jkr_t}}{4\pi r_t^2} \sin(\theta)i_\varphi. \end{cases}$$

$$(5.1.25)$$

The rest of the components are negligible. The radiated power density in the near field is then

$$P = \frac{1}{2}\mathrm{Re}(E \times H^*) = \mathrm{Re}\left(-\frac{j\eta}{2k}\left(\frac{|I|\Delta z}{4\pi}\right)^2 \frac{1}{r_t^5}\left(\sin^2(\theta)i_r - \sin(\theta)i_\theta\right)\right).$$

$$(5.1.26)$$

5.1.3 Antenna Radiation Pattern

Equation (5.1.15) can be generalized for far field radiation, where $r \gg \lambda$, to

$$E = -jk\eta \frac{e^{-jkr_t}}{4\pi r_t} F$$

$$H = -jk\eta \frac{e^{-jkr_t}}{4\pi r_t} r \times F; F = F_\theta i_\theta + F_\phi i_\phi$$

$$(5.1.27)$$

where F describes the field pattern. A real antenna is described by the power pattern P_p which is the power detected from a direction θ and ϕ with respect to the antenna axis at a fixed distance. It is defined as

$$P_\mathrm{P}(\theta, \phi) = |F_n(\theta, \phi)|^2 = |E_\theta/E_\theta(\max)|^2 \qquad (5.1.28)$$

where F_n is the normalized field pattern. Because of the value that P_p can take, often a dB representation is used.

$$P_\mathrm{P}(\theta, \phi)_\mathrm{dB} = 10 \log P_\mathrm{P}(\theta, \phi) = 20 \log|F(\theta, \phi)| = |F(\theta, \phi)|_\mathrm{dB}. \quad (5.1.29)$$

An isotropic omni-directional antenna, has a uniform radiation pattern at all angles. For infinitesimal dipole,

$$F = F_\theta i_\theta = \sin \theta$$

$$P_\mathrm{P}(\theta) = (F_n(\theta))^2 = \sin^2 \theta.$$

$$(5.1.30)$$

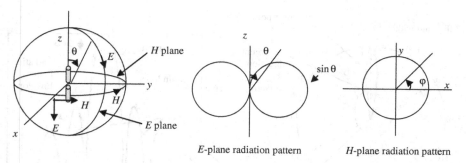

Figure 5.4 Radiation pattern for dipole

For a uniform line source pattern, the radiation pattern is

$$P_P(\theta) = (\sin(u)/u)^2; u = (\beta L/2)\cos\theta. \tag{5.1.31}$$

The radiation pattern of a half-wave dipole pattern is:

$$P_P(\theta) = \cos^2(\pi\cos\theta/2)/\sin^4\theta. \tag{5.1.32}$$

Patterns are usually plotted as a cross section of the three-dimensional surface in a plane where the planes are perpendicular cuts through the maximum lobe. Using equation (5.1.30), the radiation pattern for a dipole is represented in Figure 5.4.

Figure 5.5 shows the radiation pattern for a uniform line source. HPBW in the figure stands for half-power beamwidth. The main lobe, confined in a narrow angular range, contains most of the radiated energy. Side lobes are, most of the time, undesirable since they often constitute either a source of noise to the rest of the system or a loss of energy. The beamwidth is the width of the main lobe in a given plane. The half-power beamwidth is defined as the angular width of the main lobe between the angles at which F_n is half of its peak value or −3 dB. The antenna beamwidth is directly proportional to the wavelength and inversely proportional to the aperture diameter of the antenna. An approximation of this angle in degrees is

$$\theta_{3dB} \approx 70\lambda/d \tag{5.1.33}$$

where λ is the wavelength, and d is the antenna diameter.

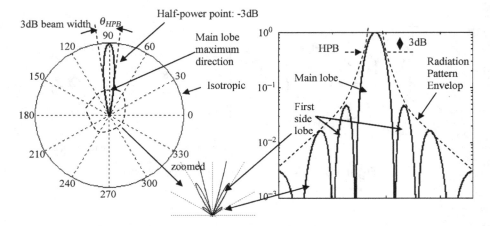

Figure 5.5 Antenna radiation pattern for a uniform line source in polar and Cartesian representation[†]

5.1.4 Directivity and Gain

The directivity of an antenna is defined as

$$D = 4\pi |F_n(\theta, \phi)|^2 \bigg/ \iint_{4\pi} |F_n(\theta, \phi)|^2 d\Omega \qquad (5.1.34)$$

where the numerator is the normalized radiation pattern and the denominator is its average value. It can also be defined in terms of power density S_r:

$$D = 4\pi S_r \bigg/ \iint_{4\pi} S_r d\Omega. \qquad (5.1.35)$$

The maximum directivity is defined as

$$D_{max} = 4\pi/\Omega_P; \Omega_P = \iint_{4\pi} |F_n(\theta, \phi)|^2 d\Omega \qquad (5.1.36)$$

[†] Figure 5.5 was generated using the following Matlab code. [†]Theta=0:1/1000:180*pi/180:u= 10.*cos (theta);p=(sin(u)./u).^2; np=p/max(p);subplot(121);semilogy(theta*180/pi,np);subplot(122);polar (theta, np).

where Ω_P is the pattern solid angle. D_{max} gives a measure of how the actual antenna performs in the direction of maximum radiation, with respect to the ideal isotropic antenna which emits the average power in all directions. A reasonable approximation of Ω_P would be to consider the product of the half-power beamwidth in both planes xz and yz for a main lobe in the z direction and no side lobes. The maximum directivity D_0 is then[6]

$$D_0 \approx 4\pi/(\beta_{xz}\beta_{yz}). \tag{5.1.37}$$

For isotropic antenna (Figure 5.6),

$$\Omega_A = 4\pi \Rightarrow D = 1. \tag{5.1.38}$$

For arbitrary direction,

$$D(\theta, \phi) = D_{max}|F_n(\theta, \phi)|^2. \tag{5.1.39}$$

The maximum directivity of an infinitesimal dipole is

$$D = 4\pi/\Omega_A$$
$$\Omega_P = \iint_{4\pi} |F_n(\theta, \phi)|^2 d\Omega = \iint_{4\pi} \sin\theta^3 d\theta d\phi \Rightarrow D = 1.5. \tag{5.1.40}$$

Directivity of half-wave dipole antenna (Figure 5.6):

$$\Omega_P = \int_0^{2\pi} \int_0^\pi (\cos(\pi\cos\theta/2)/\sin\theta)^2 \sin\theta\, d\theta\, d\phi \Rightarrow D = 1.64 \tag{5.1.41}$$

D_{max} could also be expressed in terms of antenna effective area or aperture A_e and the aperture efficiency η_a:

$$D_{max} = \frac{4\pi}{\lambda^2} A_e = \frac{4\pi}{\lambda^2} \eta_a A_P \tag{5.1.42}$$

where A_P is the antenna physical aperture.

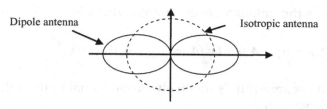

Figure 5.6 Comparison of directivity

The gain of an antenna is the ratio of the power density radiated P_r (or received) per unit solid angle by the antenna in a given direction to the power density radiated P_{ri} (or received) per unit solid angle by an isotropic antenna fed with the same power, so

$$G(\theta, \phi) = P_r/P_{ri}. \tag{5.1.43}$$

Gain is used as an intrinsic characteristic of an antenna, the higher the gain the narrower the beam is, the more energy and directivity it has, and this is the same for either receive or transmit modes.[6] The gain could also be expressed in terms of the radiation efficiency e_r which is the ratio of the radiated power to the total power supplied to the antenna:

$$G(\theta, \phi) = 4\pi e_r S_r \Big/ \int\!\!\int_{4\pi} S_r d\Omega = 4\pi r^2 S_r/P_t \tag{5.1.44}$$

where P_t is the total power supplied to the antenna. The gain of an antenna and its directivity are related by

$$G = e_r \cdot D. \tag{5.1.45}$$

For infinitesimal dipole,

$$D_{dB} = 10 \, \log G = 10 \, \log 1.5 = 1.75 \text{ dB} \tag{5.1.46}$$

The gain is maximum in the direction of maximum radiation:

$$G_{max} = 4\pi A_{eff} \frac{1}{\lambda^2} \tag{5.1.47}$$

where A_{eff} is the antenna effective area which is

$$A_{eff} = \eta A, A = \pi D^2/4 \Rightarrow G_{max} = \eta(\pi D/\lambda)^2. \tag{5.1.48}$$

The antenna beamwidth is inversely proportional to the gain relative to an isotropic radiator:

$$G \approx 4\pi/(\theta_B \phi_B) \tag{5.1.49}$$

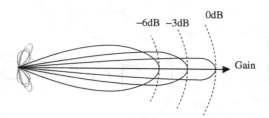

Figure 5.7 Different beam patterns for a given area

where G is the gain, θ_B is the azimuth beamwidth (rad) and ϕ_B is the elevation beamwidth (rad). The 4π in the numerator corresponds to the beam area of an isotropic radiator, and the product $\theta_B\phi_B$ in the denominator is the approximate beam area in steradians. The smaller θ_B and θ_B are, the narrower the beam and the higher the gain (Figure 5.7).

5.1.5 Antenna Polarization

Vertically and horizontally polarized antenna refers to the orientation of the current on the antenna, or the orientation of the electric part of the electromagnetic field as illustrated in Figure 5.8. Choosing a specific polarization will depend on the sensor requirements.

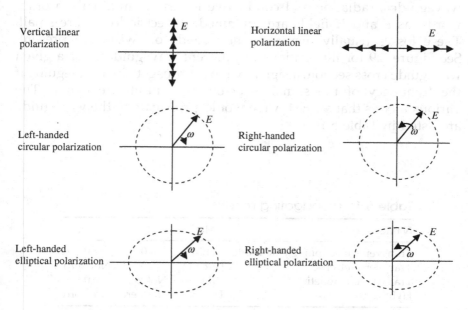

Figure 5.8 Typical antenna polarization

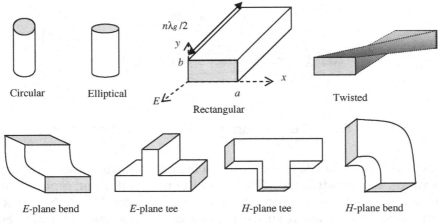

Figure 5.9 Typical waveguides

5.1.6 Waveguides

Waveguides are an alternative to conventional transmission lines which develop high losses at microwave frequencies. The cross section of the waveguide could be rectangular, elliptical or circular, depending on the design requirements. For properly designed waveguides, radiation and conductive losses are minimal. In other words, as E and H fields are contained, dielectric losses are small. These losses usually increase when bends or twists are present. See Figure 5.9 for illustrations of different waveguides. For a given waveguide cross section, a signal will not propagate in a waveguide if the frequency of the signal is below the cut-off frequency. The various modes that an EM wave would propagate in the waveguide are listed in Table 5.1.

Table 5.1 Propagating modes

Name	Mode	E_z	H_z
Transverse electromagnetic mode	TEM	0	0
Transverse electric	TE or H	0	Not zero
Transverse magnetic	TM or E	Not zero	0
Hybrid	Hybrid	Not zero	Not zero

A hollow rectangular waveguide cannot support TEM waves since there is no center conductor to satisfy Laplace's equation for potential. Multimode propagation occurs when various modes hit the waveguide walls at a different angle, and the traveled distance of each mode is different.

Waveguide properties:

$$\frac{1}{\lambda_1^2} = \frac{\varepsilon_r}{\lambda^2} = \frac{1}{\lambda_g^2} + \frac{1}{\lambda_c^2}$$

$$Z_{\text{WTE}} = \frac{377}{\sqrt{\varepsilon_r}}(\lambda_g/\lambda_1)$$

$$Z_{\text{WTM}} = \frac{377}{\sqrt{\varepsilon_r}}(\lambda_1/\lambda_g)$$

(5.1.50)

where λ_1, and λ, are successively the wavelength in the dielectric that fills the waveguide, free space, and λ_g is the guide wavelength.[1] λ_c is the cut-off wavelength, Z_{WTE} is the wave impedance for a TE mode, and Z_{WTM} is the wave impedance for a TM mode.

Group velocity v_g is used to determine the duration a signal takes to travel the length of the waveguide, while the phase velocity v_p is the rate at which a wave appears to move along a wall of a guide based on the way the phase angle varies along the guide, so

$$v_p = c/\sqrt{1 - (\lambda/2a)^2} = c/\sqrt{1 - (f_c/f)^2}$$

$$v_g = c\sqrt{1 - (f_c/f)^2}.$$

(5.1.51)

Using the phase velocity

$$\lambda_g = v_p/f = \lambda/\sqrt{1 - (\lambda/2a)^2}$$

(5.1.52)

Rectangular Waveguide

Consider the waveguide shown in Figure 5.9. The guide is filled with a material that has permittivity ε and permeability μ. For TE waves,

$$E_z = 0$$

$$\left(\partial_{xx}^2 + \partial_{yy}^2 + k_c^2\right)H_z(x, y) = 0.$$

(5.1.53)

Applying the method of separation of variables, the solution of this equation can be written as

$$H_z(x,y) = (A\cos k_x x + B\sin k_x x)(C\cos k_y y + D\sin k_y y)e^{-jk_z z}$$
$$k_z^2 = k^2 - k_c^2; \ k_c^2 = k_x^2 + k_y^2. \tag{5.1.54}$$

Since,

$$E_x = -j\omega\mu\partial_y H_z/k_c^2; \ E_y = j\omega\mu\partial_x H_z/k_c^2. \tag{5.1.55}$$

Using the boundary conditions,

$$E_x(x,0) = E_x(x,b) = 0; E_y(0,y) = E_y(a,y) = 0; \tag{5.1.56}$$

wave numbers are then

$$k_y = n\pi/b; k_x = m\pi/a; n = 0,1,2,\ldots; m = 0,1,2,\ldots \tag{5.1.57}$$

and field equations for a rectangular waveguide are therefore

$$H_z(x,y,z) = A_{mn}\cos\frac{m\pi x}{a}\cos\frac{n\pi y}{b}e^{-jk_z z}$$

$$E_y(x,y,z) = \frac{-j\omega\mu m\pi}{k_c^2 a}A_{mn}\sin\frac{m\pi x}{a}\cos\frac{n\pi y}{b}e^{-jk_z z}$$

$$H_x(x,y,z) = \frac{jk_z m\pi}{k_c^2 a}A_{mn}\sin\frac{m\pi x}{a}\cos\frac{n\pi y}{b}e^{-jk_z z}$$

$$H_y(x,y,z) = \frac{jk_z n\pi}{k_c^2 b}A_{mn}\cos\frac{m\pi x}{a}\sin\frac{n\pi y}{b}e^{-jk_z z}$$

$$E_x(x,y,z) = \frac{j\omega\mu n\pi}{k_c^2 b}A_{mn}\cos\frac{m\pi x}{a}\sin\frac{n\pi y}{b}e^{-jk_z z}.$$

The phase constant

$$k_z = \sqrt{k^2 - k_c^2} = \sqrt{k^2 - (m\pi/a)^2 - (n\pi/b)^2}. \tag{5.1.58}$$

Therefore, for $k > k_c$ the phase constant k_z is real, and thus there are one or more propagating modes. For $k < k_c$ the phase constant k_z is imaginary therefore the field components decay exponentially from where they are excited. These modes are called evanescent modes. In a waveguide if more than one mode is propagating the waveguide is said to be overmoded. Thus the cut-off frequency for any propagating mode is given by

$$k = \omega\sqrt{\mu\varepsilon}$$

$$f_{cmn} = \frac{k_c}{2\pi\sqrt{\mu\varepsilon}} = \frac{1}{2\pi\sqrt{\mu\varepsilon}}\sqrt{(m\pi/a)^2 + (n\pi/b)^2} \qquad (5.1.59)$$

where k is the wave number, ω is the angular frequency in radian s^{-1}, μ is the permeability of the medium and ε is the permittivity of the medium. The mode with the lowest cut-off frequency is called the dominant mode. Since $a > b$ for this waveguide the dominant mode is TE$_{10}$ with cut-off frequency

$$f_{c10} = \frac{1}{2a\sqrt{\mu\varepsilon}}. \qquad (5.1.60)$$

In general TE$_{nm}$ where m, n are respectively the number of variations or half cycles along b and a. The wave impedance is given by

$$Z_{TE} = \frac{E_x}{H_y} = \frac{-E_y}{H_x} = \frac{k\eta}{k_z}. \qquad (5.1.61)$$

The wave impedance is real for propagating modes while it is imaginary for evanescent modes. The guide wavelength inside the waveguide is defined as the distance between two equal phase planes along the waveguide, and is given by

$$\lambda_g = \frac{2\pi}{k_z} \succ \frac{2\pi}{k} = \lambda. \qquad (5.1.62)$$

Thus the guide wavelength is longer than the free-space wavelength. Similarly, the phase velocity of the wave inside the guide is larger than the phase velocity of the wave in free-space:

$$v_p = \omega/k_z \succ \frac{\omega}{k} = \frac{1}{\sqrt{\mu\varepsilon}} \qquad (5.1.63)$$

For instance, for a waveguide that is supposed to operate at 6 GHz where $a = 3$ cm, and $b = 1$ cm, the cut-off frequency f_{c10} of the guide wavelength, for the TE$_{10}$ mode where $\mu = 4\pi \times 10^{-7}$ and

$\varepsilon = 8.85 \times 10^{-12}$, using equation (5.1.63), is 5 GHz. The $\lambda_g = 9$ cm and the free-space wavelength, $\lambda = 6$ cm and the phase velocity inside the guide is 5.4×10^8 Hz.

For TM modes, field equations for a rectangular waveguide are

$$E_z(x,y,z) = B_{mn} \sin\frac{m\pi x}{a} \sin\frac{n\pi y}{b} e^{-jk_z z}$$

$$E_x(x,y,z) = \frac{-jk_z m\pi}{k_c^2 a} B_{mn} \cos\frac{m\pi x}{a} \sin\frac{n\pi y}{b} e^{-jk_z z}$$

$$E_y(x,y,z) = \frac{-jk_z n\pi}{k_c^2 b} B_{mn} \sin\frac{m\pi x}{a} \cos\frac{n\pi y}{b} e^{-jk_z z}$$

$$H_x(x,y,z) = \frac{j\omega\varepsilon n\pi}{k_c^2 b} B_{mn} \sin\frac{m\pi x}{a} \cos\frac{n\pi y}{b} e^{-jk_z z}$$

$$H_y(x,y,z) = \frac{-j\omega\varepsilon m\pi}{k_c^2 a} B_{mn} \cos\frac{m\pi x}{a} \sin\frac{n\pi y}{b} e^{-jk_z z}.$$

Note that for either $m = 0$ or $n = 0$ the field components are zero. Thus there are no TM_{0n} or TM_{m0} modes. The lowest order TM mode is TM_{11} and its cut-off frequency is given by

$$f_{c11} = \frac{k_c}{2\pi\sqrt{\mu\varepsilon}} = \frac{1}{2\pi\sqrt{\mu\varepsilon}} \sqrt{\left(\frac{\pi}{a}\right)^2 + \left(\frac{\pi}{b}\right)^2}. \tag{5.1.64}$$

Open Waveguide

There are many ways to transit from the waveguide to free space and vice versa. Aperture antennas radiate from a surface or an opening instead of a conducting line and are used in high frequency applications (Figure 5.10). From a rectangular aperture the normalized radiation pattern could be derived to obtain the following equation that depends on the aperture dimensions a and b, the wavelength λ, and the location of the observer θ and φ

$$F_n(\theta, \phi) = \frac{\sin^2(u\cos\phi)\sin^2(v\sin\phi)}{(u\cos\phi)^2(v\sin\phi)^2}$$
$$u = \pi a \sin\theta/\lambda$$
$$v = \pi b \sin\theta. \tag{5.1.65}$$

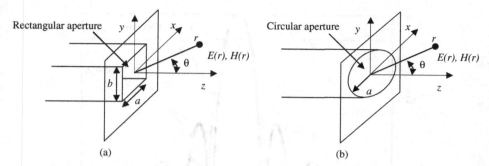

Figure 5.10 Radiation field from (a) a rectangular aperture and (b) a circular aperture

In the xz plane, i.e. where $\phi = 0$, the radiation pattern is

$$F_n = (\sin(u)/u)^2. \tag{5.1.66}$$

The half-power beamwidth in the xz and yz planes β_{xz} and β_{yz} are

$$\beta_{xz} = 0.866\lambda/a; \beta_{yz} = 0.866\lambda/b. \tag{5.1.67}$$

The maximum directivity for a rectangular aperture is

$$D_{max} = \frac{2\pi A_p}{\lambda^2} \approx 0.77 \frac{2\pi}{\beta_{xz}\beta_{yz}}. \tag{5.1.68}$$

For the circular aperture things are a bit trickier to derive. The normalized radiation pattern in the far field becomes

$$F_n(v) = 4J_1^2(v)/v^2$$
$$v = 2\pi a \sin\theta/\lambda \tag{5.1.69}$$

where J_1 is the first order of the Bessel function and a is the diameter of the aperture.[1] The corresponding directivity is

$$D_{max} = 2\pi A_p/\lambda^2 = 2\pi^2/\beta^2; \beta \approx \lambda/a \tag{5.1.70}$$

where β is the half-power beamwidth of the main lobe – see Figure 5.11 for graphs of radiation patterns of rectangular and circular apertures.

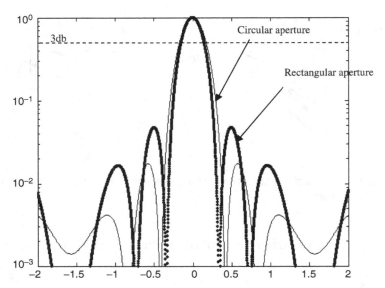

Figure 5.11 Radiation patterns of rectangular and circular apertures[†]

5.1.7 Antenna Types

An antenna is characterized by its radiation pattern, polarization, the half-power beamwidth and the frequency operating range. Table 5.2 represents some typical antennas that are used for sensing and their properties. In this section some results of directivity and other antenna characteristics will be presented. The physical characteristics of antenna are designed based on the antenna's requirement.

Horn Antennas

Between free space and a waveguide, horn antennas are used to smooth the transition between the two media. This would enhance the antenna performance. Horn antennas are often used in conjunction with reflectors that collimate EM waves into the horn. This antenna is placed at the focus of the reflector. In this section, antenna patterns and directivity are presented. The full derivations of these

---

[†] Figure 5.11 was generated using the following Matlab code. dl=3;th=−p*pi/4:1/1000:p*pi/4;ll=3; sy=pi.*dl.*sin(th);fc=bessel (1,sy);fc=fc./sy;fc=fc.*conj(fc);fr=sin(sy)./sy;fr=abs(fr.*fr);semilogy(th, fc/max(fc));hold on;semilogy(th,fr/max(fr),'.')

Table 5.2 Typical antenna pattern, polarization, half-power bandwidth (HPBW), gain G and frequency range

Antenna type	Radiation pattern	Polarization	HPBW in degrees	G in dB	Frequency range
Monopole		Linear	45	2 to 6	None
Half dipole		Linear	80	2	8 GHz
Discone		Linear	20 to 80	0 to 4	30 MHz to 3 GHz
Conical spiral		Circular	60	5 to 8	50 MHz to 18 GHz
Slot		Linear	80	6	50 MHz to 18 GHz
Guide fed slot		Linear	45 to 50	0	2 GHz to 40 GHz
Horn		Line	40	5 to 20	50 MHz to 40 GHz
Corporate feed		Variable	Number element dependent	Number element dependent	10 MHz to 10 GHz
Parabolic		Same as the feeder	1 to 10	20 to 30	400 MHz to 14 GHz

equations are beyond the scope of this book. Typical rectangular horn antennas used are shown in Figure 5.12.

An *E*-plane or *H*-plane sectoral horn can be derived from a pyramidal horn by adjusting the horn dimensions. Using basic geometry, the relationship between the horn's physical dimensions,

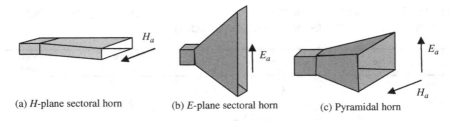

(a) H-plane sectoral horn   (b) E-plane sectoral horn   (c) Pyramidal horn

Figure 5.12 Typical rectangular horn antennas

which would be derived based on the design requirement of horn pattern, wavelength and directivity, are

$$l_E^2 = R_1^2 + (B/2)^2$$
$$\alpha_E = \tan^{-1}(B/2R_1) \qquad\qquad (5.1.71)$$
$$R_E = (B - b)\sqrt{(l_E/B)^2 - 1/4}.$$

The definitions of these terms are presented in Figure 5.13. The transverse field at the aperture of the horn is

$$E_t = E_0 \cos(\pi x/a) \exp\left(-jkR_1\left(\sqrt{1 + (y/R_1)^2} - 1\right)\right)\vec{y}$$
$$H_t = -(E_0/\eta) \cos(\pi x/a) \exp\left(-jkR_1\left(\sqrt{1 + (y/R_l)^2} - 1\right)\right)\vec{x}.$$
$$(5.1.72)$$

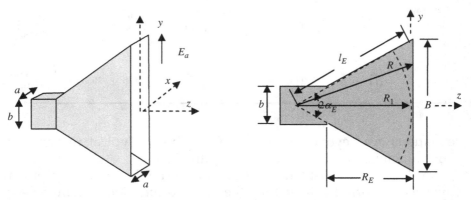

Figure 5.13 E-plane sectoral horn

The normalized patterns in the H-plane $\phi = 0$, and E-plane $\phi = \pi/2$ are

$$F_H(\theta) = \frac{1 + \cos\theta}{2} \frac{\cos[(ka/2)\sin\theta]}{1 - [(ka/2)\sin\theta]^2} \tag{5.1.73}$$

$$|F_E(\theta)| = \frac{1 + \cos\theta}{2} \left\{ \frac{[C(r_4) - C(r_3)]^2 + [S(r_4) - S(r_3)]^2}{4[C^2(2\sqrt{s}) + S^2(2\sqrt{s})]^2} \right\}^{1/2} \tag{5.1.74}$$

where

$$r_3 = 2\sqrt{s}\left[-1 - \frac{1}{4s}\left(\frac{B}{\lambda}\sin\theta\right)\right]$$

$$r_4 = 2\sqrt{s}\left[1 - \frac{1}{4s}\left(\frac{B}{\lambda}\sin\theta\right)\right]$$

$$s = \frac{B^2}{8\lambda R_2} \tag{5.1.75}$$

$$C(x) = \int_0^x \cos(\pi u^2/2)du$$

$$S(x) = \int_0^x \sin(\pi u^2/2)du.$$

$C(x)$ and $S(x)$ are Fresnel integrals.[6] The maximum phase error is $2\pi s$. The directivity for an E-plane horn antenna, which almost equals the gain, is given by

$$D_{oE} = \frac{64Al_E}{\pi\lambda B}\left(C^2\left(\frac{B}{\sqrt{2\lambda l_E}}\right) + S^2\left(\frac{B}{\sqrt{2\lambda l_E}}\right)\right). \tag{5.1.76}$$

The optimum half-power beamwidth for an E-plane horn antenna is

$$HPBW_E \approx 0.94\frac{\lambda}{B} = 54°\frac{\lambda}{B}. \tag{5.1.77}$$

Another variance of the horn antenna is the transverse field at the aperture of the H-plane horn (see Figure 5.14) is

$$E_t = E_0 \cos(\pi x/A) \exp\left(-jkR_2\left(\sqrt{1 + (x/R_2)^2} - 1\right)\right)\vec{y}$$

$$H_t = -(E_0/\eta)\cos(\pi x/A)\exp\left(-jkR_1\left(\sqrt{1 + (x/R_2)^2} - 1\right)\right)\vec{x}. \tag{5.1.78}$$

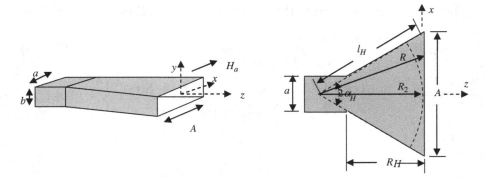

Figure 5.14 *H*-plane sectoral horn

The directivity for an *H*-plane horn antenna, which almost equals the gain, is given by

$$D_{oH} = \frac{4\pi B l_H}{\lambda A} \left((C(u) - C(v))^2 + (S(u) - S(v))^2 \right)$$

$$u = \frac{\lambda l_H + A^2}{A\sqrt{2\lambda l_H}}$$

$$v = \frac{\lambda l_H - A^2}{A\sqrt{2\lambda l_H}}.$$

(5.1.79)

The optimum half-power beamwidth for an *H*-plane horn antenna is

$$HPBW_H \approx 1.36\frac{\lambda}{A} = 78°\frac{\lambda}{A}.$$

(5.1.80)

The transverse field at the aperture of the pyramidal horn (Figure 5.15) is

$$E_t = E_0 \cos(\pi x/A) \exp\left(-j\frac{1}{2}k(x^2/R_2 + y^2/R_1)\right)\vec{y}$$

$$H_t = -(E_0/\eta) \cos(\pi x/A) \exp\left(-j\frac{1}{2}k(x^2/R_2 + y^2/R_1)\right)\vec{x}.$$

(5.1.81)

The directivity for a pyramidal horn antenna, which almost equals the gain, is given by

$$D_{op} = \frac{\pi\lambda^2}{32AB}D_{oE}D_{oH}.$$

(5.1.82)

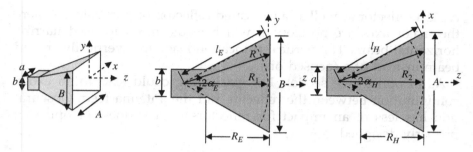

Figure 5.15 Pyramidal horn antenna

The design of the horn will be based on the required radiation pattern, directivity and HPBW for a given frequency of operation. This leads to the geometrical properties of the horn.

Reflectors

The perfect analogies to radio frequency (RF) reflectors are mirrors in optics. As in optics, in order to collect the maximum radiation and focus it to the entrance of an RF circuit, reflectors are used to bounce either the incoming radiation or the emitted one from a feeder, such as a horn antenna. This would allow, for example, the adjustment of the width of the beam and also the side-lobe level. The placement of the feeder depends on the design of the reflector and its focal point. The shape of the reflector depends on the wavelength, feeder, swath and other factors. Antenna focusing can be accomplished with parabolic reflectors which restrict the direction of waves detected by the dipole or horn antenna at the focus. Figure 5.16 shows a

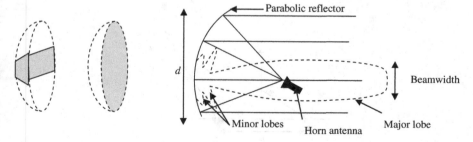

Figure 5.16 Parabolic reflector, truncated parabolic reflector and parabolic reflector with a horn antenna located at the focus

typical reflector as well as a truncated reflector, or paraboloid, where the major axes are horizontal, which leads into a focused narrow horizontal beam. The vertical truncation leads to a vertically spread beam instead of a focused one.

As in optics (see section 1.6), packaging could lead to an exotic configuration between the reflector and the antenna feed. Obscuration has less of an impact for reflectors than it does for optics, if properly designed.

5.2 Phased Arrays

An array of antennas is a conglomeration of more than one antenna arranged in such a way as to enhance the power and directivity. This occurs by wisely designing the antennas and the RF front end to make the interference between each individual antennas constructive rather than destructive in a specific direction. In the transmitting mode each individual element radiates precisely in phase to produce wave that move forward in order to interfere constructively to produce a strong narrow beam. In the receiving mode the power received by each element is the sum of the received power scattered by target P from all the transmit elements. Virtually any radiation pattern in the far field could be accomplished by meticulously choosing some of these parameters: spacing between the antenna, the type of each individual antenna, the number and placement of each antenna in regard to its neighbors and the start time of transmission or reception of the signal. Phased array deals mainly with adjusting the phase between individual antennas which enable, for example, beam steering.

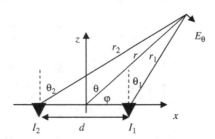

Figure 5.17 Two dipoles separated by a distance d

5.2.1 Simple Array of Two Antennas

To analyze the beam pattern of a phased array, a two-antenna element is used first. Consider two identical parallel dipoles separated by a distance d (Figure 5.17). The dipole currents have the same amplitude and total phase difference α.

Using equations (5.1.16) the electrical far field components are

$$E_{\theta_1} \approx jk\eta I_0 \Delta z \frac{e^{-jkr_1+j\alpha/2}}{4\pi r_1} \sin \theta_1$$

$$E_{\theta_2} \approx j\eta k I_0 \Delta z \frac{e^{-jkr_2-j\alpha/2}}{4\pi r_2} \sin \theta_2 \qquad (5.2.1)$$

$$\eta = \sqrt{\frac{\mu}{\varepsilon}}$$

where α is the phase between the two identical antennas and referenced at the center. At long distance:

$$r \gg d; \theta_1 \approx \theta_2 \approx \theta; r_1 \approx r - \frac{d}{2}\cos\varphi; r_2 \approx r + \frac{d}{2}\cos\varphi. \qquad (5.2.2)$$

Therefore the total field which is the superposition of both fields is

$$E = E_1 + E_2 \Rightarrow |E|^2 = |E_0|^2 \cdot A_F$$

$$E_0 = \eta \frac{jk I_0 \Delta z \sin(\theta)}{4\pi r} e^{-jkr} \qquad (5.2.3)$$

$$A_F = 4\cos^2\left(\frac{kd\cos(\varphi) + \alpha}{2}\right)$$

where A_F is the array factor. E_0 is the equivalent field of the dipole located at the center of the array. Figure 5.18 is a representation of the array factor for different phase and antenna spacing.

5.2.2 Linear Antenna Array

For a uniform, identical n-element antenna array, which is shared by the same current of constant amplitude but with an increasing equal phase α, the total electrical field at a point of interest will be derived in a way similar to the case of two antennas. It is important to maintain the same length for each transition to control phase shifts

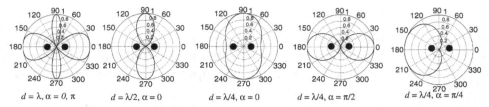

$d = \lambda, \alpha = 0, \pi$ $d = \lambda/2, \alpha = 0$ $d = \lambda/4, \alpha = 0$ $d = \lambda/4, \alpha = \pi/2$ $d = \lambda/4, \alpha = \pi/4$

Figure 5.18 Array factor for various wavelengths, spacing ratio and phase[†]

between each antenna. Figure 5.19 shows an n-element antenna array.

Here it is assumed that there are no mutual couplings among the elements, i.e. the input impedance of each of the antenna elements is not affected by the presence of other antenna elements. Following the same steps as for the two dipoles case, the n dipoles at long distance are

$$E = E_1 + E_2 + \cdots + E_n$$

$$E = E_0 e^{-jkr} + E_0 e^{-jk(r-d\cos\psi)} e^{j\alpha} + \cdots + E_0 e^{-jk(r-d(n-1)\cos\psi)} e^{j(n-1)\alpha}$$

$$E = E_0 e^{-jkr} \frac{1 - e^{jn\varphi}}{1 - e^{j\varphi}}$$

$$\varphi = kd\cos\psi + \alpha = \frac{2\pi d\cos\psi}{\lambda} + \alpha. \qquad (5.2.4)$$

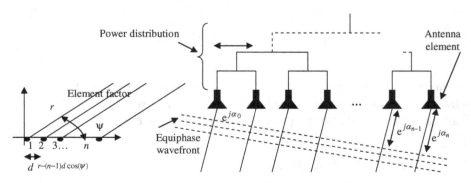

Figure 5.19 n-element antenna array

[†]Figure 5.18 was generated using the following Matlab code. $d$lambda=1/4;alfa=pi/4;f=0:0.001:9;
y=2*cos(pi*$d$lambda*cos(f)+ alfa*pi/2);y=y/max(y);y=y.*y;y;polar(f,y);

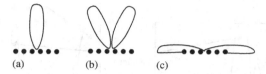

Figure 5.20 (a) Broadside array, (b) intermediate array and (c) end-fire array

The magnitude of the electric field is given by

$$|E|^2 = |E_0|^2 A_F$$

$$A_F = \frac{\sin^2(n\varphi/2)}{\sin^2(\varphi/2)}$$

(5.2.5)

$$A_{Fn} = \frac{\sin^2(n\varphi/2)}{n^2 \sin^2(\varphi/2)}.$$

Equation (5.2.5) shows that the A_{Fn}, the normalized array factor or the group pattern, has a maximum when φ is a multiple of 2π. The maximum where $\varphi = 0$ is called the main lobe. When the main lobe is perpendicular to the line of the array, i.e. $\psi = \pi/2$, the array is called a broadside array and is illustrated in Figure 5.20. When the main lobe is aligned along the array, i.e. $\psi = 0$, the array is called an endfire array and is also illustrated in Figure 5.20. The rest of the lobes are called grating lobes, which are undesirable. Grating lobes are additional major lobes with high intensity. Grating lobes, often unwanted, occur when

$$\varphi = 2m\pi = 2\pi d \cos \psi / \lambda + \alpha.$$

(5.2.6)

Therefore, for a linear antenna array, grating lobes can be suppressed by judicially choosing the spacing between elements. See Figure 5.21 for graphs of linear array patterns with and without grating lobes.

For n dipoles, a total of $(n-1)$ lobes and $(n-2)$ side lobes could be formed. Figure 5.22 shows that as the number of dipoles increases, the main beam narrows and the side-lobes' level decreases. The limitation on the number of antenna per array is power, size and weight. The complexity of the circuits that drives each antenna also increases. The directivity of the phased linear array is proportional to

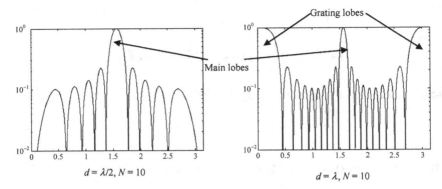

Figure 5.21 Linear array pattern with and without grating lobes[†]

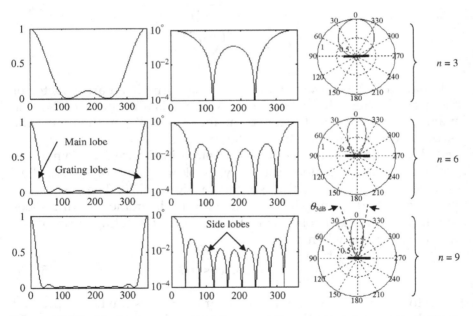

Figure 5.22 Various representations of the main lobe and grating lobes, and radiation pattern for linear arrays of various numbers of elements[‡]

[†]Figure 5.21 was generated using the following Matlab code. dl=1/2;$n = 10$;$p = 40$;th=$-$p*pi/4:0.0001:p*pi/4;sy=2.*pi.*dl.*cos(th);f=abs (sin(n*sy/2)./sin(sy/2))/n;semilogy(th,f);axis([0 pi 0 1]).
[‡]Figure 5.22 was generated using the following Matlab code. f=0:0.001:2*pi;n=9;y=sin(n*f/2)./(n*sin(f/2));$y = y.*$ y;polar(f,y);semilogy (f*180/pi,y).

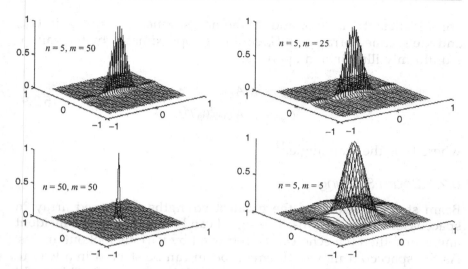

Figure 5.23 Three-dimensional representation of $n \times m$ rectangular array[†]

the number of antenna, and the half beamwidth θ_{3dB} is

$$D_0 \approx n$$
$$\theta_{3dB} = 2\cos^{-1}(1 - \lambda/nd). \tag{5.2.7}$$

As part of a requirement, the beamwidth and directivity are specified. For instance for $n = 16$, $D_{0dB} = 12$ dB and $\theta_{3dB} = 7$ deg.

5.2.3 Two-dimensional Antenna Array

In the case of a two-dimensional rectangular array, the radiation pattern could be approximated as the product of the patterns of the two planes that contain the principle axes of the antenna:

$$F(\theta, \phi) = F(\theta)F(\phi) = \frac{\sin^2(n\pi d \sin\theta/\lambda)\sin^2(m\pi d \sin\phi/\lambda)}{\sin^2(\pi d \sin\theta/\lambda)\sin^2(\pi d \sin\phi/\lambda)}. \tag{5.2.8}$$

One can see in Figure 5.23 that by choosing n and m judicially, the antenna pattern directivity and the side lobes can be designed well.

[†]Figure 5.23 was generated using the following Matlab code. clear hold off, plot3(1,1,1) d=1; l=2;n= 50;m=50;i=0;for theta=−pi/4;1/25:pi/4;i=i+1;j=0; for phi=−pi/4:1/25:pi/4;j=j+ 1;ut=pi*d*sin (theta)/l;up=pi*d*sin(phi)/l;p(j)=(sin(n*ut)*sin(m*up)/(sin(ut)*sin(up)))^2;pphi(j)=phi;ttheta(j)=theta; end;ma(i)=max(p);hold on;plot3 (ttheta,pphi,p/2.5e + 006);end;max(ma).

For a sufficiently large n and m, the non-scanned antenna gain G_{ns}, and the scanned array gain G_s could be approximated by the gain of a uniformly illuminated aperture:

$$G_{ns} \approx 4\pi A/\lambda^2$$
$$G_s \approx 4\pi A \cos\theta_0/\lambda^2$$
(5.2.9)

where θ_0 is the scan angle.[11]

5.2.4 Beam Steering

Beam steering is one of the greatest strengths of phased array in spacebased radar systems. The signal can be sent at a given incident angle, and then the echo can be retrieved by various lookout angles. As the spacecraft moves, the main beam can be steered in a way to compensate for the forward movement of the spacecraft. This could be very useful to direct a particular area of interest. Figure 5.24 shows that for a given phase tuned by phase shifters, the signal and its echo can be stirred to maximize the SNR, and enhance the directivity for a given target.

In the event where $\alpha_1 = \alpha_2 = \ldots = \alpha_n$, the main beam will be broadside to the array. The direction of the main beam will be θ_0 if the relative phase difference between elements is

$$\alpha = 2\pi d \sin\theta_0/\lambda.$$
(5.2.10)

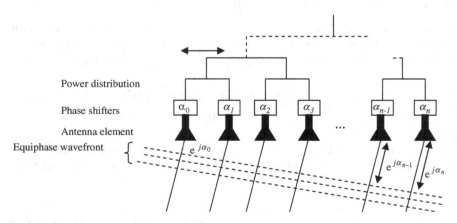

Figure 5.24 Beam steering using phase shifters in a corporate feed at each radiating element[11]

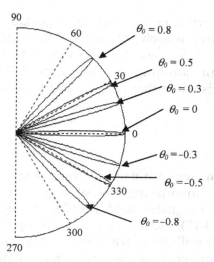

Figure 5.25 Steering the beam by changing the phase, for this case the array has 20 antennas[†]

The steered radiation pattern is then

$$f(\theta) = \frac{|E_a|^2}{N^2} = \frac{\sin^2(N\pi d(\sin\theta - \sin\theta_0)/\lambda)}{N^2 \sin^2(\pi d(\sin\theta - \sin\theta_0)/\lambda)}. \tag{5.2.11}$$

Figure 5.25 shows the main beam direction for a given θ_0. The half-power beamwidth of a scanned array can be approximated by the following formula (not valid near endfire):

$$\theta_{3dB} \approx 0.886\lambda/Nd\cos\theta_0. \tag{5.2.12}$$

Similar to phased array, grating lobes are undesirable since they are considered as a loss of power and a source of noise in the measured signal, therefore they should be suppressed by carefully choosing the spacing between the antennas. Equation (5.2.10) shows that grating lobes will occur at

$$\frac{\pi d}{\lambda}(\sin\theta_g - \sin\theta_0) = \pm n\pi, \tag{5.2.13}$$

[†]Figure 5.25 was generted using the following Matlab code. th0=−0.9;dl=.5;n=20;p=4;th=−p* pi/4:1/1000:p*pi/4; sy=pi.*dl.* (sin(th) − sin(th0));f=((sin(n*sy))/n).^2; hold on;polar(th,f).

The near field pattern is rather complex especially if the coupling or the interaction between each antenna is accounted for. Typically, in order to quantify these effects, one element is excited and the response of the surrounding ones is analyzed. This could identify any impedance mismatch that could result in spurious lobes.

5.3 Radiometers

Spacebased radiometers are used for hydrology, agriculture, ocean monitoring, storm monitoring, climatology, atmospheric temperature profiles and magnetic field profiles. Using the atmospheric window (section 4.17), clouds or dusts are opaque for IR sensors but transparent for millimeter wave receivers such as radiometers. A microwave radiometer is a passive sensor where the operating wavelength ranges between 20 and 100 cm. Using the Rayleigh–Jean approximation where $hc \ll \lambda kT$, the emission of an object in the microwave region is linearly dependent on temperature:

$$M_\lambda = \varepsilon M_{bb\lambda} = \frac{2\pi ck}{\lambda^4}\varepsilon T = \frac{2\pi ck}{\lambda^4}T_b \Rightarrow T_b = \varepsilon T \qquad (5.3.1)$$

where T_b is the brightness temperature, M_λ is the excitance or spectral radiant emittance, which is expressed in terms of energy per unit frequency, ε is the emissivity and T is the physical temperature of the radiating element. Since all bodies radiate less than a black body at the same temperature, T_b is defined such that the brightness of the body is the same as a black body at that temperature. For instance, the brightness temperature of a body at 300 K with an emissivity of 0.9 is 270 K. Alternatively the power density per unit Hz, M_v, is often used:

$$\nu = c/\lambda \Rightarrow d\nu = -cd\lambda/\lambda^2$$
$$|M_\lambda d\lambda| = |M_v d\nu| \Rightarrow M_v = 2\pi kT^2/\lambda^2. \qquad (5.3.2)$$

The spectral brightness B_f which is the radiation intensity per unit area and bandwidth, for a Lambertian surface (see section 4.2.2), is

$$B_f = M_v/\pi = 2kT/\lambda^2. \qquad (5.3.3)$$

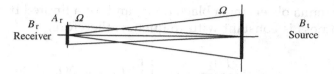

Figure 5.26 Power received from a source

5.3.1 Power–temperature Correspondence for Antennas

The power intercepted by the antenna with aperture A_r as illustrated in Figure 5.26 is

$$P = S_t A_r = F_t A_r / R^2 \tag{5.3.4}$$

where P is the received power in W, S_t is the target power density in $\mathrm{Wm^{-2}}$, A_r is the receiver antenna aperture in $\mathrm{m^2}$, and F_t is the radiation intensity in W steradian^{-1}. The brightness is defined as the radiated power per unit solid angle per unit area:

$$B = F_t / A_t. \tag{5.3.5}$$

Using the relationship between the solid angle and the source aperture A_t, and substituting it into the power equation

$$\Omega_t = A_t / R^2 \Rightarrow P = B A_r \Omega_t \tag{5.3.6}$$

the differential power received by the antenna for a differential solid angle is

$$\partial P = A_r \cdot B \cdot F_n \partial \Omega, \tag{5.3.7}$$

where B is the source brightness as a function of a solid angle and F_n is the normalized radiation pattern of the antenna as a function of the solid angle. The power incident on the antenna in terms of the brightness of the source of the radiation and the gain pattern of the antenna is obtained by integrating over 4π steradians and over the frequency band f_1 to f_2:

$$P = \frac{A_r}{2} \int_{f_1}^{f_2} \int \int_{4\pi} B \cdot F_n \partial \Omega \partial f. \tag{5.3.8}$$

For an antenna observing a black body, and for a limited bandwidth, the brightness is constant with frequency, therefore,

$$P_{bb} = \frac{kT(f_2 - f_1)A_r}{\lambda^2} \int\int_{4\pi} F_n(\theta, \phi)\partial\Omega. \qquad (5.3.9)$$

Since the pattern solid angle Ω_p is

$$\Omega_p = \int\int_{4\pi} F_n(\theta, \phi)\partial\Omega = \frac{\lambda^2}{A_r} \qquad (5.3.10)$$

therefore the equation of radiometry is

$$P_{bb} = kT(f_2 - f_1) = kT\beta \qquad (5.3.11)$$

where β is the bandwidth. This relationship shows that the detected power is independent of the antenna gain, the distance from the radiating target and the temperature of the antenna structure.[12] It also shows that the detected power is proportional to temperature.

5.3.2 Remote Temperature Measurement using Radiometry

In radiometry an antenna measures an apparent temperature T_{AP} of a matched resistor with noise power output of kT_{AP} at the antenna interface, where k is Boltzmann's constant. This replaces the received power as the measure of signal strength. T_{AP} depends on the path and the object of interest. Therefore, a general formulation of the T_{AP} is

$$T_{AP} = T_{UP} + (T_B + T_{SC})/L_A \qquad (5.3.12)$$

where T_B is the brightness temperature from target emissions and T_{SC} is the scatter temperature, which is radiation reflected from the target into the main lobe of the antenna.[12] This might include components of the atmospheric downwelling temperature T_{DN}. L_A is the atmospheric attenuation of the total radiation. T_{UP} is the upward emission temperature from the entire atmosphere between the target and the spacebased radiometer as presented in Figure 5.27.

For instance, a spacebased radiometer that operates at 36 GHz with a bandwidth of 1 GHz that is pointing at an area that is at 30°C

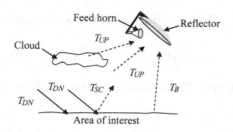

Figure 5.27 Radiometer scene temperatures

and has an average emissivity of $\varepsilon = 0.95$, the received power is computed using equation (5.3.12). For a total loss through the atmosphere of 1.5 dB, and a downwelling brightness temperature of 40 K. Assuming that the $T_{UP} \sim T_{DN}$, T_{AP} is then

$$T_{AP} = 40 + (0.95 \times 303 + (1 - 0.95) \times 40)/1.42 = 244\,\text{K}. \qquad (5.3.13)$$

Using equation (5.3.11), the power is

$$P = 10\log_{10}(1.38 \times 10^{-23} \times 244 \times 10^9) = -114.72\,\text{dBW} = -84.72\,\text{dBm}. \qquad (5.3.14)$$

Thus far the antenna was assumed to be lossless. In reality some of the power incident on the antenna is absorbed, and some is dissipated or radiated. The antenna noise temperature, which is the apparent temperature measured at the antenna output T_{AO}, is

$$T_{AO} = \eta_1 T_A + (1 - \eta_1)T_p \qquad (5.3.15)$$

where η_1 is the radiation efficiency of the antenna, T_A is the scene temperature measured by a lossless antenna, and T_P is the physical temperature of the antenna. In this case the efficiency η_1 could be seen as the antenna reflectivity. The apparent noise power is

$$P_{AO} = kT_{AO}\beta = k(\eta_1 T_A + (1 - \eta_1)T_p)\beta \qquad (5.3.16)$$

where β is the bandwidth and k is Boltzmann's constant. The transmission line and receiver combination impact the measured temperature. Indeed, for a transmission line with a loss factor L that is connected to a low-noise receiver at temperature T_r, the equivalent input noise power P_c from this combination is then

$$P_c = kT_c\beta = k\big((L-1)T_p + LT_r\big)\beta. \tag{5.3.17}$$

All things considered, the total-system equivalent input power P_S is then

$$P_S = P_{AO} + P_c \Rightarrow T_A = \frac{P_S - k\big((L-\eta_1)T_p + LT_r\big)\beta}{\eta_1 k \Delta f}. \tag{5.3.18}$$

In radiometry the parameter of interest for measurement is T_A. In order to obtain that value all the other parameters of this equation need to be measured. P_S is proportional to the receiver output voltage and therefore is an actual measured value. Knowing all the variables of equation (5.3.18) at the time of measurement presents a challenge, especially for a harsh, space-like environment. The actual measurement of the target of interest depends exclusively on the knowledge of these variables. Therefore the temperature of a well defined source, a reference temperature that is located as close as possible to the receiver, could be used. Often, a calibration source that is a known reference source for the sensor is measured between the actual measurement of the area or target of interest. The switching between the target and the calibration source should be fast enough to ensure that during the actual measurement the radiometer environment did not change.[12]

For a typical radiometer antenna the half-power 3dB beamwidth is

$$\theta_{3dB} \approx 70\lambda/D \tag{5.3.19}$$

where D is the diameter of the antenna in m, λ is the wavelength in m, and θ_{3dB} is in degrees. If the area of interest has similar emissivity, the size of the antenna footprint does not affect the terrain brightness. T_B is then

$$T_B = T_{BG}(1 - F) + T_{BT}F$$
$$F = A_T/A = A_T/R\theta \tag{5.3.20}$$

where T_{BG} is the ground brightness temperature, T_{BT} is the target brightness temperature, A_T is the target area, A is the antenna footprint, R is the altitude of the spacebased radiometer and F is the beam fill ratio.

Since the receiver is not at 0 K, even if there is no external source, the radiometer will produce an output signal which is defined as

$$P_N = k T_{sys} \beta \cdot G \qquad (5.3.21)$$

where P_N is the output thermal noise power, T_{sys} is the system noise temperature, k is Boltzmann's constant, β is the system bandwidth in Hz and G is the system power gain.

$$T_{sys} = T_0(f - 1); f = SNR_{in}/SNR_{out} \qquad (5.3.22)$$

where f is the noise factor of the receiver and T_0 is the classic temperature of the system. Obviously, the cooler the system is the lower this noise power is. However, rather than working hard in cooling the system, and because of the presence of other kinds of noise sources that vary continuously during measurements, the approach should be to calibrate the sensor between measurements. This could be done using a reference load or source and switching between the target measurement and the reference, then extracting the actual measurement of interest. The following section explores one of these methods of calibration.

5.3.3 Dicke Radiometer

A Dicke radiometer relies on switching the measurement between the target and a reference load. The switching is commanded by a switch driver that is tuned to the square law generator (Figure 5.28). A Dicke switch could be a chopper wheel, a semiconductor diode or a ferrite circulator, depending on the switching rate which should be higher than twice the bandwidth of the intermediate filter (IF). The switching time for pin diodes is an order of magnitude smaller than the on for the ferrite.

The output of the square law detector is synchronously detected[12] to produce an output that is proportional to the difference between T_{cal} and T_A:

$$\Delta T_{min} = (T_A + 2T_{sys} + T_{cal}) \sqrt{ \frac{1}{\beta_{IF}\tau} + \left(\frac{\Delta G}{G} \cdot \frac{T_A - T_{cal}}{T_A + 2T_{sys} + T_{cal}} \right)^2 }.$$

$$(5.3.23)$$

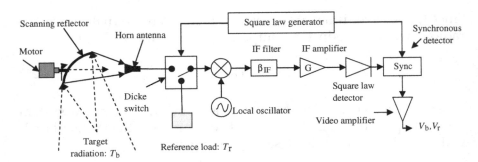

Figure 5.28 Typical RF front end of microwave radiometer using a Dicke switch and a scanning reflector

Another scheme, called the correlation scheme, is sometimes used instead of the Dicke scheme. This scheme does not have moving parts, contrary to the Dicke switch. It uses high modulation frequency, an improvement in sensitivity over a single mixer and a capability of sideband separation. The drawbacks of this approach are that it uses a larger number of components and that a bandwidth filter is required for good phase and amplitude matching.

Most radiometers employ a square law detector which produces an output signal proportional to the square of the input current. The detector response for this case is

$$\langle V \rangle = a_2 \langle I^2 \rangle + a_4 \langle I^4 \rangle + \cdots \tag{5.3.24}$$

Using time-averaged voltage, the relationship between the voltage and T_S is

$$\langle V \rangle = a + bT_S(1 + \mu T_S)$$
$$\mu = 3a_4 k \beta G / a_2 \tag{5.3.25}$$

where k is the Boltzmann's constant, β is the bandwidth, T is the amplifier temperature, T_S is the radiometric temperature and G is the intermediate frequency amplifier gain (Figure 5.29).

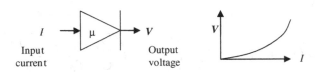

Figure 5.29 Square law detector response

where a and b are functions of the amplifier temperature, its gain and the bandwidth. These two parameters are exclusive to the receiver and its environment. They are subject to change continuously because of the nature of the spacecraft environment. Therefore, two-point radiometer calibration measurements, both the cold space and a warm target, are important to conduct between measurements in order to eliminate a and b from $\langle V \rangle$ so that the brightness temperature T_b viewed by the radiometer could be approximated by

$$T_b = T_1 + (V_b - V_1)\left(\frac{T_2 - T_1}{V_2 - V_1}\right) + \mu\left(\frac{T_2 - T_1}{V_2 - V_1}\right)^2 (V_b - V_1)(V_b - V_2),$$

(5.3.26)

where V_b is the output voltage from the target of interest and 1 and 2 correspond to the first target and the second target.[12] Because of the nature of the spacecraft environment, often one of the targets corresponds to the cold space and the other is an onboard warm target.

5.3.4 Radiometric Sensitivity

The radiometric sensitivity ΔT is the smallest detectable variation in the antenna noise temperature T_A and T_p by a radiometer. When the input of a radiometer is band-limited white noise, ΔT could be derived by analyzing its output. For a typical radiometer front end as represented in Figure 5.30, a rough estimate of the sensitivity is then derived.

For a rectangular IF, with a bandwidth β_{IF}, the magnitude P_{IF} is

$$P_{IF} = \frac{kTG_{IF}}{2}$$

(5.3.27)

where G_{IF} is the gain of the IF. The output of the square law detector includes a DC component P_{DC} which is the average value of the input voltage, input power P_{SYS}, and an AC component which is

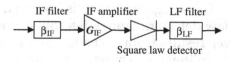

Figure 5.30 Typical radiometer front end

the statistical uncertainty associated with the measured input voltage which has a magnitude P_{AC} and width β_{IF}:

$$P_{DC} = \left(\frac{kTG}{2}\right)^2 \beta_{IF}^2$$

$$P_{AC} = \left(\frac{kTG}{2}\right)^2 \beta_{IF}.$$

(5.3.28)

For the final filtered output signal, a low-pass filter with a bandwidth β_{LF} that is much smaller than β_{IF} filters the AC component. While the DC component remains unchanged, the total AC power is then

$$P_{AC} = 2\left(\frac{kTG}{2}\right)^2 \beta_{IF}\beta_{LF}.$$

(5.3.29)

The temperature change ΔT, radiometer sensitivity, can be measured by V_{AC}. V_{DC} will measure the sum of the system and antenna temperatures. Since the two ratios are the same, therefore

$$\frac{\Delta T}{T_A + T_{sys}} = \frac{V_{AC}}{V_{DC}} = \sqrt{\frac{P_{AC}}{P_{DC}}} \Rightarrow \Delta T = (T_A + T_{sys})\sqrt{\frac{2\beta_{LF}}{\beta_{IF}}}.$$

(5.3.30)

For a low-pass filter with a time constant τ then:

$$\beta_{LF} = \tau/2 \Rightarrow \Delta T = (T_A + T_{sys})/\sqrt{\beta_{IF}\tau}.$$

(5.3.31)

The sensitivity equation shows that the measurable temperature difference ΔT decreases when the system temperature T_{sys} decreases.[12] This could be achieved, for example, by cooling the receiver. ΔT decreases by increasing the IF bandwidth β_{IF} or by increasing the integration time τ. The integration time is bounded by the observation time. Square law detectors cannot determine whether the change in the signal power results from a change in T_A or in a change in the gain of the IF amplifier G_{IF}. Therefore the minimum temperature change detectable is

$$\Delta T_{\min} = (T_A + T_{\text{sys}}) \left[\frac{1}{\beta_{\text{IF}}\tau} + \left(\frac{\Delta G}{G_{\text{IF}}}\right)^2 \right]^{1/2}.$$
(5.3.32)

5.4 Aqua: AMSR-E

The advanced microwave scanning radiometer for the Earth (AMSR-E) observing system was part of the Aqua satellite (Figure 5.31). It is a 324 kg passive microwave radiometer system that was developed and built by the Japan Aerospace Exploration Agency (JAXA). It is part of the Aqua spacecraft as shown in Figure 5.31. It measures geophysical variables such as precipitation rate, cloud water, water vapor, sea surface winds, sea surface temperature, sea ice concentration and soil moisture. It measures brightness temperatures at 6.9 GHz, 10.7 GHz, 18.7 GHz, 23.8 GHz, 36.5 GHz and 89.0 GHz. Spatial resolutions of the individual measurements vary from 5.4 km at 89 GHz to 56 km at 6.9 GHz.

The 1.6 m diameter reflector and the feedhorn arrays (Figure 5.32), where the incoming radiation is focused, are mounted on a drum.

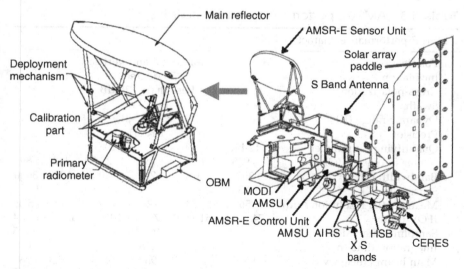

Figure 5.31 Aqua satellite and the AMSR-E (Courtesy of NASA, USA and JAXA, Japan)

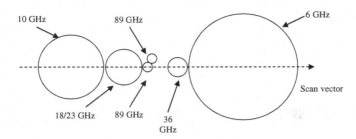

Figure 5.32 The six feedhorn location (Courtesy of JAXA, Japan)

The drum contains the RF and digital subsystem mechanical scanning subsystem and power subsystem. The whole system rotates about the axis of the drum at 40 rpm which allows a swath width of 1445 km. Table 5.3 presents the main physical and performance characteristics of the AMSR-E.

There are mainly three sources of errors in AMSR-E, estimated to be 0.66 K at 100 K and 0.68 K at 250 K. The first source of error is the warm load reference error resulting from the inaccuracy of the thermistors. The second one is the inhomogeneous temperature

Table 5.3 AMSR-E performance characteristics[8]

AMSR-E performance characteristics						
Polarization	Horizontal and vertical					
Inclination	55 degrees					
Cross-polarization	Less than −20 dB					
Swath	1445 km					
Dynamic range (K)	2.7 to 340					
Precision	1 K (1)					
Quantifying bit number	12-bit		10-bit			
Center frequency (GHz)	6.925	10.65	18.7	23.8	36.5	89
Bandwidth (MHz)	350	100	200	400	1000	3000
Sensitivity (K)	0.3		0.6		1.1	
Mean spatial resolution (km)	56	38	21	24	12	5.4
IFOV (km)	74 × 43	51 × 30	27 × 16	31 × 18	14 × 8	6 × 4
Sampling interval (km)	10 × 10				5 × 5	
Integration time (msec)	2.6				1.3	
Main beam efficiency (%)	95.3	95	96.3	96.4	95.3	96
Beamwidth (degrees)	2.2	1.4	0.8	0.9	0.4	0.18
Stability	80 arcsec/sec/axis for roll and pitch; N/A for yaw					
Pointing accuracy	600 arcsec/axis for roll and pitch; N/A for yaw					

distribution of the load area, and from the feed horn couplings: The third source is the cold load reference error which results mainly from the coupling between the cold sky reflector and the feed horn. The radiometer non-linearity error results from the square law detector and the temperature dependent gain drift of the receiver electronics. Both the cold load including the cold-sky mirror and reflector, and the warm load are mounted off-axis on a transfer assembly shaft that does not rotate with the drum assembly.

5.5 SeaSat: SMMR

The scanning multichannel microwave radiometer (SMMR) is a passive microwave radiometer that was first part of the SeaSat spacecraft (Figure 5.33), then used in the Nimbus series of space-

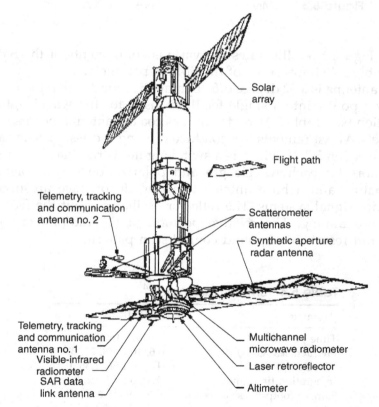

Figure 5.33 SeaSat spacecraft (Courtesy of NASA/JPL/Caltech)

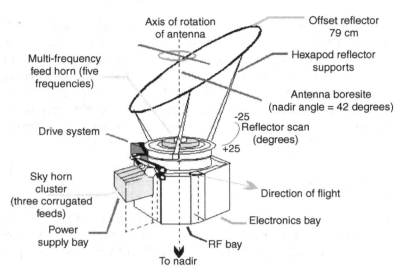

Figure 5.34 SMMR (Courtesy of NASA/JPL/Caltech)

crafts. Figure 5.34 illustrates the various components of the SMMR and Table 5.4 shows some of the SMMR parameters.

The antenna is a 42 degree offset parabolic reflector that focuses the received power into a single feedhorn used for the wavelengths of operation (see Table 5.4), and provides coaxial antenna beams for all channels. Measurements are conducted using a classic Dicke radio-meter (section 5.3.3), which is a switching network of latching ferrite circulators for each wavelength and polarization. An ambient RF termination and a horn antenna viewing deep space are used as reference signal systems. The reflector oscillates about the feedhorn axis for scanning. The relative position of the radiometer to the spacecraft results in a ground conical scan pattern.

Table 5.4 SMMR parameters[4]

Parameter	SMMR (Nimbus-7)
Time period	1978 to 1987
Frequencies (GHz)	6.6, 10.7, 18, 21, 37
Wavelength (cm)	4.54, 2.8, 1.66, 1.36, 0.81
Integration time (ms)	126, 62, 62, 62, 30
Sample footprint sizes (km)	148 × 95 (6.6 GHz)
	27 × 18 (37 GHz)

5.6 Envisat: MWR

Envisat (illustrated in Figure 5.35) is a spacecraft that was built by the European Space Agency (ESA). It was an Earth polar satellite orbiting at 800 km. The microwave radiometer (MWR, Figure 5.36) is one of the instruments that Envisat had on board. The 24 kg MWR instrument is a nadir pointing instrument. Its FOV is 20 km diameter. It uses the Dicke radiometer principle (refer to section 5.3.3), and operates at vertically polarized 23.8 GHz and 36.5 GHz. The Dicke method reduces the impact of the gain and noise temperature instabilities on the measurement (Table 5.5).

There are two operational modes. The first one is the nominal Dicke operation mode. In this mode a continuous comparison between the temperature and an internal reference load is conducted. The second mode is the calibration mode. This mode is a two-point calibration. The first point is a cold reference which is a measurement of the deep cold space. It is conducted using a sky-horn pointing into

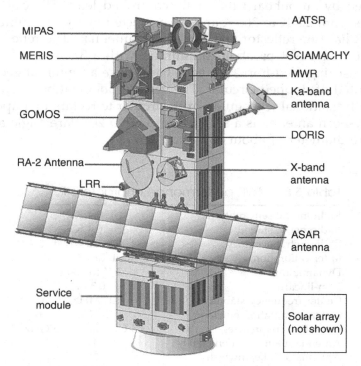

Figure 5.35 Envisat (Courtesy of European Space Agency ESA Remote Sensing Missions ©)

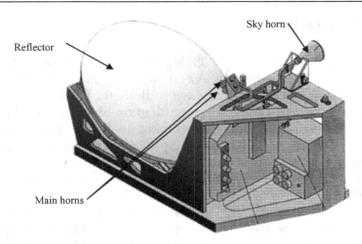

Figure 5.36 MWR sensor description. (Courtesy of European Space Agency ESA Remote Sensing Missions ©)

the cold deep space. The second reference point is a hot reference provided by an onboard thermally controlled load. The calibration range is from 2.7 K to 300 K and is conducted every few minutes.

MWR has one reflector and three horn antennas. The reflector is a 60 cm chord offset parabolic aluminum with a 35 cm focal length. The measuring feed-horns for each band have an internal geometry constituted of stepped circular. The vertical polarization is extracted using a transitional waveguide from circular to rectangular aperture. The sky-horn antenna is a scalar horn with a 20° flare angle and an aperture diameter of about 80 mm.

Table 5.5 MWR performance characteristics[5]

Radiometric sensitivity	0.4 K
Radiometric stability	0.4 K
Radiometric accuracy	1 K at Ta = 300 K
(after calibration)	<3 K at Ta = 85 ÷ 330 K
Dynamic range	3 K to 330 K
Non-linearity	0.35 K
Center frequency stability	<0.2MHz/°C
Antenna radiation efficiency	97%
Operating frequencies	23.8 GHz and 36.5 GHz
Antenna main beam efficiency	94%
Antenna 3dB beamwidth	1.5°
Orbit	800 km
Operational power	18 W

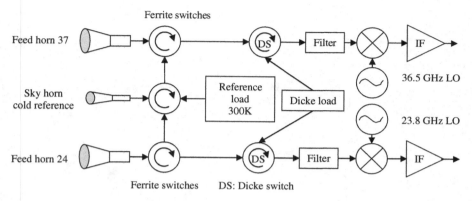

Figure 5.37 MWR: RF front end[5]

Figure 5.37 shows the RF front end. The microwave radiation is focused on the feed horn using the offset parabolic reflector. Ferrite circulators alternate the measurement between ground measurement cold space measurement and the reference load. It is then routed through a Dicke switch assembly to a down converter which shifts the K- and Ka-Band signals to an appropriate IF range for processing and interpretation.

Bibliography

1. Bahria, P. and Bahi, I. (1984) *Millimeter Wave Engineering and Applications,* John Wiley and Sons.
2. Barton, D. K. (1988) *Modern Radar System Analysis,* Artech House, Inc.
3. Blake, R. (2002) *Electronic Communication Systems,* Delmar.
4. Elachi, C. (1987) *Introduction to the Physics and Techniques of Remote Sensing,* John Wiley & Sons.
5. ESA (2004) *Envisat RA2-MWR Product Handbook.*
6. Lee, K. F. (1984) *Principles of Antenna Theory,* John Wiley and Sons Ltd.
7. Mailloux, R. J. (1994) *Phased Array Antenna Handbook,* Artech House.
8. NASDA (2003) *AMSR-E Data Users Handbook, 2nd edition.*
9. Reedy, E. K. and Eaves, J. L. (1987) *Principles of Modern Radars,* Van Nostrand Reinhold Company.
10. Rees, W. G. (1990) *Physical Principles of Remote Sensing,* Cambridge University Press.
11. Skolnik, M. I. (1980) *Introduction to Radar Systems,* McGraw-Hill.
12. Ulaby, F. T., Moore, R.K. and Fung, A. K. (1981) *Microwave Remote Sensing: Active and Passive,* Volume I, Addison-Wesley Publishing Company, Inc.
13. Ulaby, F. T., Moore, R. K. and Fung, A. K. (1986) *Microwave Remote Sensing: Active and Passive,* Volume III, Addison-Wesley Publishing Company.

6

Spacebased Radar Sensors

6.1 Radar Introduction

The word radar stands for radio detection and ranging. It includes all active EM sensors that rely on a returned signal, i.e. echoed signal, which results from the interaction of the incoming signal and the target. Radars are used to measure target information such as range, radial velocity, angular direction and target size. There are many applications of radar in space. The military uses it to monitor ground, ocean and space activities in a region of interest. Civilians use it to monitor ocean activities and weather patterns, and to manage disaster areas.

Typically a radar system (Figure 6.1) is composed of a wave generator, that could be a discrete digital synthesizer (DDS), which creates a baseband signal. This generator creates a phase and amplitude of continuous waves (CW) or linear frequency modulation (LFM) type of waves. An up-converter mixes a baseband signal up to the desired transmit frequency. This is done using mixer circuits and frequency multipliers. A transmitter amplifies the signal. An antenna directs the transmitted signal. The echoed signal is collected by the same antenna that runs it to a low-noise amplifier (LNA) which should be as close as possible to the front end. The echoed signal is then mixed with a reference signal generated by a local oscillator to down-convert the signal to intermediate frequencies (IF) in order to

Spacecraft Sensors.   Mohamed M. Abid
© 2005 John Wiley & Sons, Ltd

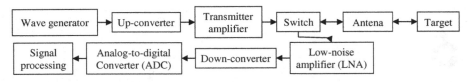

Figure 6.1 Typical radar system

fit into the bandwidth of the analog-to-digital converter (ADC). The received signal is subject to a wide variety of noise sources and interferences that are either related to the receiver itself, such as the internal thermal noise, or related to the surrounding environment of the target or the receiver, such as the energy received from galactic sources, neighboring radar systems, neighboring communication systems, radar jammers and clutter (energy reflection from surrounding objects like buildings and hills). Therefore filters and amplifiers are placed in various steps to further enhance the SNR. A signal processing unit processes the echoed signal for target identification or range measurement. The processing level depends on the mission resources. Often raw data or low level processed data is down-linked to the ground for further processing since extensive processing time is often needed.

In the following sections, an introduction to radar will be presented with an emphasis on key parameters that will be used in spacebased sensor design. Typical applications of radars such as altimeters, synthetic aperture radar (SAR), and ground penetrating radar will be described.

6.1.1 Overview

A radiation source that is tuned to the target characteristics is radiated by an antenna that is designed to maximize the energy on a target. The impact on the target of interest results in an isotropic, random or polarized radiation. Some of that reflected power in the form of an EM wave reaches a receiver antenna that converts it to an electrical signal that is filtered, amplified and processed. The signal undergoes a variety of changes in amplitude and form. Figure 6.2 shows the various power values during this signal round-trip. This will be explored in the following section. After the travel is complete, the signal is dramatically weakened, but still could be detected in the receiver.

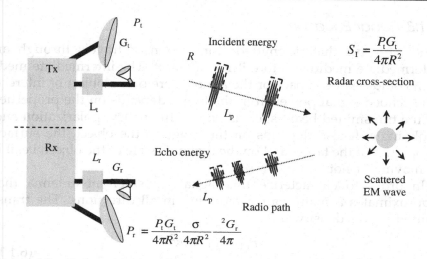

Figure 6.2 Concept of radar operation – two antennas are used for illustration purposes but usually only one antenna is used

Parameters used in Figure 6.2 are P_t, G_t, L_t, P_r, G_r, L_r, and L_p, representing transmit power, transmit gain, transmitter losses, received power, receiver gain, receiver losses and propagation losses respectively. The transmitter is Tx and the receiver is Rx.

6.1.2 Frequency Bands

Spacebased radar uses RF signals that range from the K band to the P band (see Figure 4.2 and Table 4.1). The choice of the appropriate frequency of operation is intrinsic to the mission design and target of interest nature, resolution and attenuation. Typically, once the frequency and bandwidth are determined, a certification has to be granted from the International Telecommunication Union (ITU). The purpose of this certification is to ensure that there is no impact of the frequency chosen on existing radar, communication or any other sources that may use similar frequencies. Interferences with other RF sources could be very harmful to both systems, especially for radars that are in the receiving mode. In this mode, amplifiers and filters are both active. An unanticipated high power signal in the appropriate frequency reaching the receiver can dramatically damage the receiver.

6.1.3 Radar Equation

An EM wave that is emitted from a sensor travels through an intermediate medium before it hits a target. The intermediate medium could be open space or the atmosphere of the object of interest. The echoed signal power from the target depends on the properties of the transmitted beam such as power, frequency, polarization and look angle. It also depends on the range to the object, the surface properties of the target and on the dynamic state of the object, i.e. if it is moving or not.

In order to characterize this signal, consider an antenna that approximates a point source radiating in all directions. The transmitted power density is

$$P_{dt} = P_t/4\pi R^2$$
$$P_{target} = \sigma P_t/4\pi R^2,$$

(6.1.1)

where P_{dt} is the power density at the target due to an isotropic source, P_t is the emitted power from an omni-directional antenna, 4π is the solid angle about a point and R is the range (Figure 6.3). P_{target} is the power intercepted at the target, in W; σ is the radar cross section of the target in m^2, which is a characteristic of the target. Notice the R^2 decay of the power density, which means that the further the target is the lower the power density. Treating the target as a Lambertian 'reflector' that transmits or echoes a portion of radiation, the scattered or reflected radiation by the target is emitted equally in all directions. The reflected power density P_{dr} at the radar receiver is

$$P_{dr} = \sigma P_t/\left(4\pi R^2\right)^2.$$

(6.1.2)

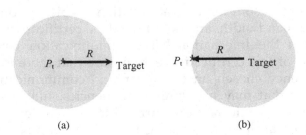

(a)   (b)

Figure 6.3 (a) The target receives a portion of the transmitted signal; (b) the target scatters the intercepted radiation and acts as a reflector that radiates to the initial transmitter

The power drops by $1/R^2$ twice; once on the way to the target, and once on the way back from the target. Let A_r be the effective aperture, area, of the unity gain radar receiving antenna. The power at the antenna terminal P_r will be given by

$$\left. \begin{array}{l} P_r = \sigma P_t A_r / (4\pi R^2) \\ A_r = \lambda^2 / 4\pi \end{array} \right\} \Rightarrow P_r = \sigma P_t \lambda^2 / (4\pi)^3 R^4. \qquad (6.1.3)$$

Most real antennas are directional, i.e. they emit radiation into a fixed solid angle. The concentration of power into a relatively small solid angle is described by the antenna gain G_t. Therefore the power density from a directive antenna is

$$EIRP = P_t G_t \Rightarrow P_{dr} = EIRP\sigma / (4\pi R^2)^2 \Rightarrow P_r = P_t G_t \sigma A_r / (4\pi R^2)^2,$$
$$(6.1.4)$$

where $EIRP$ stands for the equivalent isotropically radiated power or the effective isotropic radiated power, which is the power that would be radiated by an isotropic antenna to give the same radiated power as the given antenna. A radar antenna receives a portion of the re-radiated signal. Antenna theory explains the relationship between the gain of a lossless antenna as related to its aperture A or effective area by

$$G_t = 4\pi A_t / \lambda^2$$
$$G_r = 4\pi A_r / \lambda^2 \qquad (6.1.5)$$
$$P_r = \sigma P_t \lambda^2 G_t G_r / (4\pi)^3 R^4,$$

where A_t is the effective area of the transmitting antenna and A_r is the effective area of the receiving antenna. Often the same radar antenna is used for both transmitting and receiving, especially since the transmitted signal and the received signal are separated by the flight time between the target and the transmitter. Therefore, the gains of the antenna for the receiving and transmitting modes are the same and P_r becomes

$$P_r = \sigma P_t \lambda^2 G^2 / (4\pi)^3 R^4 = \sigma P_t A^2 / (4\pi \lambda^2 R^4). \qquad (6.1.6)$$

There are many sources of dissipation, or noise, between the sensor and the target that would lead to a degradation or attenuation of the signal. They could eventually be captured together and denoted by L where $L < 1$. The reduced received power is thus

$$S = P_t G^2 \lambda^2 \sigma L / (4\pi)^3 R^4. \tag{6.1.7}$$

6.1.4 The Radar Range Equation

There are many factors that affect radar range. On the transmitter side, range is affected by transmitter power, pulse length, antenna gain and its beamwidth and pattern. On the receiver side it is affected by the receiver threshold, its noise figure and the fact that the receiver antenna is the same as the transmitter antenna. The target cross section and the wavelength also affect as well the radar range. Using the radar equation, the maximum detectible range R_{max} is

$$R_{max} = \left(\sigma P_t \lambda^2 G^2 / (4\pi)^3 P_{s\,min} \right)^{1/4}. \tag{6.1.8}$$

The maximum range R_{max} is determined by the minimum detectable echo signal power $P_{s\,min}$ required by the receiver to detect and distinguish the target. Since adding is easier than multiplying square and cubic parameters, in addition to the wide range that a parameter can take, the detection range and most parameters in the radar area are expressed in decibels (dB). Using the decibel equivalent for each parameter, the previous equation becomes

$$40 \log_{10} R_{max} = P_{tdB} + 2G_{dB} + 10 \log_{10} \left(\lambda^2 / (4\pi)^3 \right) + \sigma_{dB} - P_{s\,min\,dB}. \tag{6.1.9}$$

For the design requirements, this equation by itself is not sufficient. An *SNR* requirement is needed in order to compute R_{max}.

Detection Range in Terms of Output SNR

In the case, where only thermal noise is present, using equation (3.1.1), equation (6.1.8) becomes

$$R_{max} = \left(\frac{\sigma P_t \lambda^2 G^2}{(4\pi)^3 fk T_{sys} \beta_n SNR_{o\,min}} \right)^{1/4}. \tag{6.1.10}$$

This equation shows the various impacts of temperature, bandwidth and *SNR*. The received signal from a target is usually down-converted to lower frequencies for processing. In order to maximize the signal-to-noise ratio where both the signal and noise are present, the receiver frequency response is matched to the transmitted one. This is the so-called matched filter. The maximum value of the SNR that could be achieved:

$$(S_p/N)_{\text{out}} = SNR_{\text{out}} = 2E/N_o \qquad (6.1.11)$$

where S_p is the peak instantaneous signal power seen during the matched filter response to a pulse, N is the average noise power, E is the received signal energy and N_o is the single-sided noise power density.[7] Since the received energy is the product of the received power and the pulse duration τ, and the noise power density is the ratio of the received power by the bandwidth, equation (6.1.11) becomes

$$SNR_{\text{out}} = (S/N)_{\text{in}} \beta_{\text{IF}} \tau = SNR_{\text{in}} \beta_{\text{IF}} \tau. \qquad (6.1.12)$$

To optimize the matched filter, the receiver bandwidth should be equal to the transmitted bandwidth. Also, since the β_{IF} is small compared with the centre frequency, the peak power is twice the average power in the received pulse. Therefore the minimum received signal in terms of *SNR* is

$$S_{\text{min}} = SNR_{\text{out}} N/\beta\tau = SNR_{\text{out}} kT \cdot f/\tau. \qquad (6.1.13)$$

Using the range equation, the maximum range becomes

$$R_{\text{max}} = \left[\frac{P_t G^2 \lambda^2 \sigma \tau}{(4\pi)^3 kT_o SNR_{\text{out}}} \right]^{1/4}. \qquad (6.1.14)$$

The implementation of the matched filter is usually associated with some loss in *SNR*. This loss depends on the type of input signal and the filter shape. Table 6.1 shows some typical losses. It is common to use $\beta\tau$ of 1 and a loss of 0.5 dB.

In a more general form, the *SNR* could be written in the following way:

$$SNR = \left(\frac{P_t G^2 \lambda^2}{(4\pi)^3 \rho^4 L_t L_r f} \right) \left(\frac{\sigma}{kT_n B_n} \right) \left(\frac{\rho\lambda}{L_a} \frac{c\tau}{2\sin\theta} \right) \qquad (6.1.15)$$

Table 6.1 Input signal and filter shape impact on losses in SNR[2]

Input signal	Filter shape	$\tau\beta$	Loss in SNR (dB)
Rectangular	Rectangular	1.37	0.85
Rectangular	Gaussian	0.72	0.49
Gaussian	Rectangular	0.72	0.39
Gaussian	Gaussian	0.44	0

where P_t is the peak transmit power, G is the one way antenna gain, L_t is the transmit system loss, L_r is the receive system loss, f is the operating noise figure, k is the Boltzmann's constant, T_n is noise temperature, B_n is bandwidth, τ is the pulse length and L_a is the antenna length. This equation accounts for many losses that could be found from emitting the signal to receiving the echoed signal.[7]

There are many ways to improve the SNR for a fixed radar frequency. The implementation of each method has its repercussion on the system where a trade study can optimize the option. For instance, increasing the antenna gain by increasing the size of the antenna affects the total mass and volume of the payload. The transmitted power is limited by the power available from the space-craft. The pulse length could be increased or the bandwidth could be decreased but the resolution could decay and impact power. Another approach would be to use a different orbit to reduce R but the impact of choosing a lower orbit could decrease the coverage of the space-craft, which may require multiple orbits to cover a full area of interest. Also in lower orbit orbital perturbations increase which may impact some sensors. By using shorter wavelengths the system becomes more susceptible to atmospheric dispersions. A typical method to maintain similar radar resolution while increasing the radar pulse length is to modulate the signal. Many modulation schemes are used for this purpose. Using FM chirps is the most popular scheme. This will be explored in section 6.1.7.

Radar Pulse

Target range is measured by noting the time difference between the transmission of a pulse and the reception of an echo (Figure 6.4),

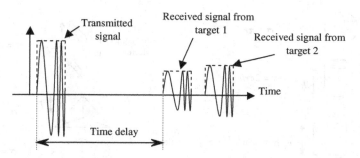

Figure 6.4 Pulse and its echoes

which is the detected received signal that exceeds a certain threshold. The range is given by

$$R = c\Delta t/2 \qquad (6.1.16)$$

where Δt is the time difference or time delay and c is the speed of EM waves.

In order to be able to distinguish a target's echo from each pulse without mixing responses from secondary pulses, targets should lie within the range from 0 to $T/2$. The corresponding range is

$$R_{\text{unamb}} = c/2f_{\text{r}} \qquad (6.1.17)$$

where R_{unamb} is the range for unambiguity and f_{r} is the pulse repetition frequency (PRF). The inverse of the PRF is called the pulse repetition interval (PRI, see Figure 6.5).

The range resolution δR, which is the minimum range separation of two targets, is a function of the transmitted signal bandwidth β.

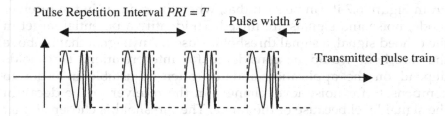

Figure 6.5 Transmitted and received pulse train, i.e. a modulated signal

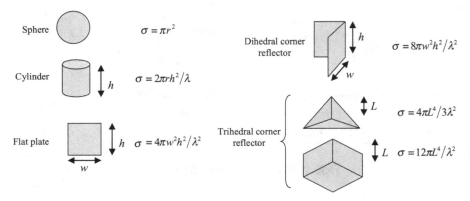

Figure 6.6 Typical RCS values

For a point target, β could be approximated by the inverse of the pulse width τ. Therefore,

$$\delta R = c\tau/2. \tag{6.1.18}$$

6.1.5 Radar Cross Section

The radar cross section (RCS) σ, expressed in m^2, is the effective area ascribed by the target which is different from the geometrical area. The RCS is related to the size of the target, its configuration, the viewing angle and the wavelength of the radar operating frequency. It is an intrinsic part of the echo signal which is a measure of the ratio of reflected power per unit solid angle in the direction of the radar to the power density intercepted by the target. RCS gauges the target reflection of an incident radar signal in the direction of the radar. Typical RCS values are shown Figure 6.6.

6.1.6 False Alarm

From Figure 6.7 it can be seen that, when the radar is in the receiver mode, noise and signals are mixed. To identify a potential target in the echoed signal, a signal threshold is used. Any signal that is above the threshold would be considered for interpretation. Thresholds depend on the application and are often adjustable in order to compensate for noise level changes in the receiver or the decay in the signal level because of the range. The threshold is chosen so that the probability of false alarm, P_{fa}, which occurs when the noise exceeds the threshold, is low. A trade-off between the P_{fa} and range

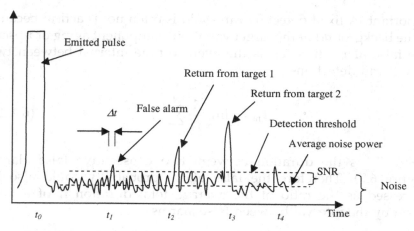

Figure 6.7 False alarm duration and detection threshold

or target identification determines the appropriate threshold to be used. This approach does not eliminate the possibility of missing a target. This occurs when the target signal is below the noise level and the threshold.

The receiver noise entering the IF filters is well described using a Gaussian probability density function (PDF)

$$p(v) = \frac{1}{\sqrt{2\pi\psi_o}} \exp \frac{-v^2}{2\psi_o} \tag{6.1.19}$$

where $p(v)dv$ is the probability of finding the noise voltage v between v and $v + dv$. ψ_o is the variance of the noise voltage. If the Gaussian noise is passed through a narrow band filter, the PDF of the envelope of the noise voltage is

$$p(A_f) = \frac{A_f}{\psi_o} \exp \frac{-A_f^2}{2\psi_o} \tag{6.1.20}$$

where A_f is the amplitude of the envelope at the filter output. The probability of false alarm P_{fa} is determined by integrating the PDF:

$$P_{fa} = \int_{V_t}^{\infty} \frac{R}{\psi_o} \exp \frac{-R^2}{2\psi_o} dR = \exp \frac{-V_t^2}{2\psi_o} \tag{6.1.21}$$

where V_t is the voltage threshold. This equation shows that P_{fa} is very sensitive to V_t. Therefore, a careful selection of V_t is very important. A fixed detection threshold is often not practical because of the background of the target variation, component aging or upsets. The false alarm time T_{fa} is the average time interval between two false alarm detections:

$$T_{\text{fa}} = \lim_{N \to \infty} \frac{1}{N} \sum_{k=1}^{N} T_k \qquad (6.1.22)$$

where T_k is the duration between two consecutive false alarms (Figure 6.7). Therefore the false alarm probability could also be expressed as the ratio of the average total duration t_k of a false alarm by the interval between false alarms:

$$P_{\text{fa}} = \langle t_k \rangle_{\text{ave}} / \langle T_k \rangle_{\text{ave}} = 1/T_{\text{fa}}\beta. \qquad (6.1.23)$$

Since typically the average duration of a noise pulse $\Delta t_k \approx 1/\beta$ where β is the bandwidth, therefore,

$$T_{\text{fa}} = \frac{1}{\beta_{\text{IF}}} \exp \frac{V_t^2}{2\psi_o}, \qquad (6.1.24)$$

where β_{IF} is the intermediate frequency bandwidth. The probability of detection, P_{D}, occurs when the echoed signal is received from an actual target of interest. Generally the target detection, P_{D}, and false alarm, P_{fa}, probabilities are expressed in terms of the output signal-to-noise ratio. The empirical relationship between P_{D}, P_{fa}, and the numerical SNR,[10] not in dB, is

$$SNR = a + 0.12ab + 1.7b$$
$$a = \ln(0.62/P_{\text{fa}}) \qquad (6.1.25)$$
$$b = \ln(P_{\text{D}}/(1 - P_{\text{D}})).$$

The selection of P_{fa} and P_{D} is determined by the application, but typically P_{fa} varies between 10^{-7} and 10^{-3}, and P_{D} varies between 0.1 and 0.9.

6.1.7 Doppler Radars and the Doppler Effect

The Doppler effect occurs whenever there is a frequency shift resulting from the moving of the target or the radar platform, a moving spacecraft or an incident wave, for example. This allows the detection of a target, and the measurement of its radial velocity and range used in remote sensing, docking, rendezvous, descent and other maneuvers. There are many radar systems such as pulsed radar continuous wave (CW) radar, and FM-CW radar that will be explored in the following sections.

Doppler CW Radar

The Doppler CW radar emits a continuous, as opposed to pulsating, unmodulated wave in a given frequency and measures the echoed signal frequency shift known as the Doppler frequency shift. The difference between the transmitted frequency and the echoed frequency is known as the beat frequency f_b.

The total number of wavelengths of the two-way path between the radar and the target is given by $2R/\lambda$. Since a wavelength corresponds to an angular excursion of 2π radians, the returned echo will have a phase difference given by[10]

$$\left. \begin{array}{l} \lambda = c/f_t \\ \phi = 4\pi R/\lambda \end{array} \right\} \Rightarrow \phi = 4\pi R f_t/c \qquad (6.1.26)$$

where f_t is the radar transmitted frequency and c is the speed of radio waves. For a moving target, R and f are functions of time and f changes because of the Doppler effect:

$$\phi(t) = 4\pi R(t) f_t/c. \qquad (6.1.27)$$

Differentiating R and f with respect to time, the equation of the Doppler frequency is

$$\partial_t \phi(t) = \frac{4\pi R(t) f_t}{c} \partial_t R$$

$$\partial_t \phi(t) = \omega_r = 2\pi f_d \qquad (6.1.28)$$

$$\partial_t R = v_r \Rightarrow f_d = \frac{2 v_r f_t}{c}$$

where f_d is the Doppler frequency, v_r is the target relative velocity, f_t is the radar transmit frequency, $c \sim 3 \cdot 10^8$ ms^{-1}.

In Figure 6.8, $E_{r,a}$ is the echo signal from an approaching target, $E_{r,r}$ is the echo signal from a receding target and $\omega_d = 2\pi f_d$, which is the

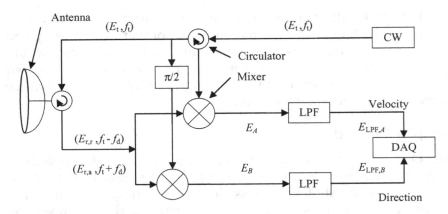

Figure 6.8 Radial velocity detection[10]

frequency of targets on the beam centerline.

$$E_t = E_0 \cos(\omega_0 t)$$
$$E_{r,a} = kE_0 \cos((\omega_0 + \omega_d)t) \qquad (6.1.29)$$
$$E_{r,r} = kE_0 \cos((\omega_0 - \omega_d)t)$$

$$E_A \propto \cos(\omega_d t) + \cos((2\omega_0 + \omega_d)t) \Rightarrow E_{LPF,A} \propto \frac{1}{2}\cos(\omega_d t) \qquad (6.1.30)$$

$$E_B \propto \cos(\omega_d t + \pi/2) + \cos((2\omega_0 + \omega_d)t - \pi/2)$$

$$\Rightarrow E_{LPF,B} \propto \frac{1}{2}\cos(\omega_d t + \pi/2). \qquad (6.1.31)$$

The delay in this figure is important for digital processing, where one end would be the imaginary component of the returned signal, and the other end would be the real component. Signal processing could include Fourier transforms or correlation. The low pass filter (LPF) is used to remove the upper side band from mixing.

LFM Radar Systems

CW radar systems do not provide the range to the target. Using frequency modulation techniques, range information can be derived. Frequency and phase modulation are of particular interest because they have the characteristic of widening the bandwidth for high resolution while increasing the radiated energy for higher pulse width. In order to minimize noise the receiver's bandwidth should be equal to the inverse of the transmitted pulse, otherwise noise is

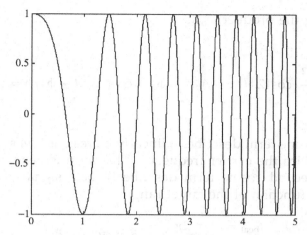

Figure 6.9 Typical chirp

introduced for larger bandwidths, and ringing for smaller bandwidths. The most popular FM used in range determination is the linear frequency modulation (LFM) known as chirp (Figure 6.9). Chirp is used when the pulse is required to have a high peak power to exceed the noise present in the system. Chirp is often designed to achieve a large time–bandwidth product. The resulting phase is quadratic in time, which has a linear derivative. In most active sensors, the frequency is modulated in a linear manner with time

$$\omega_b = A_b t \tag{6.1.32}$$

where A_b is the chirp rate. Substituting into the standard equation for FM, the following result is obtained:

$$v_{fm}(t) = A_c \cos\left(\omega_c t + A_b \int_{-\infty}^{t} t dt\right) = A_c \cos\left(\omega_c t + \frac{A_b}{2} t^2\right). \tag{6.1.33}$$

All of the returned signal is mixed with a locally created replica of the transmitted signal. The transmit signal will be shifted from that of the received signal because of the round trip time τ

$$v_{fm}(t - \tau) = A_c \cos\left(\omega_c(t - \tau) + A_b(t - \tau)^2/2\right). \tag{6.1.34}$$

---

† Figure 6.9 was generated using the following Matlab code. t=0:0.001:0.5; wc=1; Ab=3; y=cos((Ab*t/2).^2+wc*t);y=y/max(y); plot(t,y)

Calculating the product, v_{out}:

$$v_{out}(t) = v_{fm}(t - \tau)v_{fm}(t)$$

$$= \frac{A_c^2}{2}\left(\cos(2\phi t/\tau + A_b t^2 - \phi) + \cos(A_b \tau.t + \phi)\right) \phi = \omega_c \tau - \frac{A_b}{2}\tau^2$$

$$(6.1.35)$$

The first cosine term describes a linearly increasing FM signal; chirp, at about twice the carrier frequency. This term is generally filtered out. The second cosine term describes a beat signal at frequency which is a function of the target range:

$$\left.\begin{array}{l} f_{beat} = A_b\tau/2\pi \\ \tau = 2R/c \end{array}\right\} \Rightarrow R = \pi c f_{beat}/A_b, \qquad (6.1.36)$$

(Figure 6.10 shows f_{beat}). The signal frequency is directly proportional to the delay time τ, and hence is directly proportional to the distance from a reference point. The reference point may be a near edge of interest or the centre of a scene.[16] Figure 6.11 illustrates the full sequence of chirp transmit, received echoed signal and de-chirp.

The amplitude went from 1 to the square root of $\beta\tau$. Generating the chirp and the de-chirp could be done using the same filter that would properly match the waveform. The frequency rate of change is

$$\dot{f}_0 = \Delta f/\Delta t. \qquad (6.1.37)$$

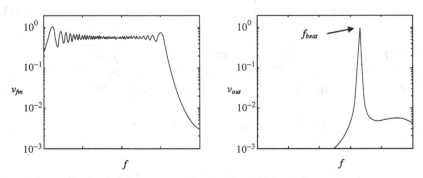

Figure 6.10 Spectrum of a linear FM signal

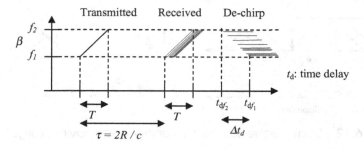

Figure 6.11 Frequency–time relationship of a linear chirp

If the frequency is being increased linearly at the rate f_0 in Hz, then during the time $\Delta t = 2R/c$ the transmitted frequency f_t would have increased by $\Delta f = f_0 \, \Delta t$. Therefore,

$$f_b = \Delta f = f_0 \, 2R/c \Rightarrow R = cf_b/2f_0 \,. \qquad (6.1.38)$$

Triangular Modulation

Since the frequency cannot be increased indefinitely and the target may not be stationary, a modulating signal waveform is used which could be either a pulsating signal or a triangular signal, linearly increasing and decreasing. For the triangular signal, with a frequency f_m, and the peak-to-peak frequency deviation $2\Delta f$, then the transmitted frequency is linearly increasing and decreasing by the rate

$$\dot{f}_0 = \Delta f/\Delta t = 4f_m \Delta f; f_m = 1/T, \qquad (6.1.39)$$

where f_m is the modulating frequency. Using equation (6.1.38), the range as a function of the beat frequency is

$$R = cf_b/(8f_m \Delta f). \qquad (6.1.40)$$

When the target is moving the Doppler effect takes place and the frequency is shifted accordingly as presented in Figure 6.12. For a moving target or moving platform, the beat frequency would include an additional component due to the Doppler frequency shift. The positive and negative frequency slopes will be: $f_{up} = f_r - f_d$; $f_{down} = f_r + f_d$.

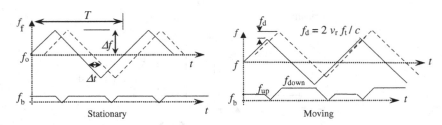

Figure 6.12 Beat frequency for stationary and moving target waveform[10]

Moving Target Indication (MTI) and Pulsed Doppler Radar System

While Doppler CW radar systems provide the relative velocity of a moving object, pulsed radar systems provide range information. MTI and pulsed Doppler radar systems are used to detect moving targets in a severely cluttered environment. While both types measure range and velocity, MTI radars are optimized for unambiguous range and velocity measurements.

Figure 6.13 shows that the pulse width is tuned based on the measurement that the sensor is designed for. Figure 6.14 shows that to remove the ambiguity in the measurement, two PRFs can be used since different PRFs have different range ambiguities. The target's range could be determined by finding the intersection of range for different PRFs.

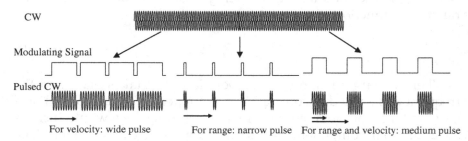

Figure 6.13 Modulating signal depends on the measurement of interest[10]

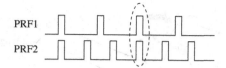

Figure 6.14 Two different PRFs can remove the ambiguity

A Doppler Perspective

Whether the target is moving and the platform is steady or vice versa, as illustrated in Figure 6.15, there is a Doppler shift. For instance, when the target enters the forward edge of the beam the received signal will be shifted in frequency relative to the transmitted one by f_d:

$$
\left.
\begin{aligned}
v_r &= v\cos\theta \\
\theta &\approx \frac{vt}{R}
\end{aligned}
\right\}
\Rightarrow f_d(t) = \frac{2v_r}{\lambda} = \frac{2v}{\lambda}\cos\left(\frac{vt}{R}\right)
\qquad (6.1.41)
$$

where v_r is the radial velocity. Targets ahead of the radar will have positive Doppler shifts, and those behind the radar have negative Doppler shifts.

6.1.8 Resolution

Radar imaging resolution is determined by the pulse length (which affects the range resolution), the beamwidth (which affects the azimuth resolution), the size of the antenna and the orbit of the platform.

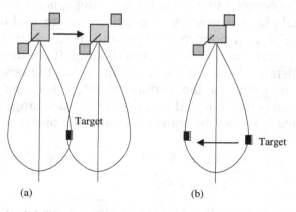

(a)   (b)

Figure 6.15 In (a) the target is steady and the platform moves; in (b)

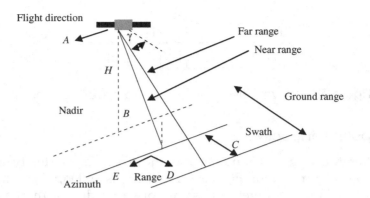

Figure 6.16   Geometry[16]

For a radar application in remote sensing, the platform travels forward in the flight direction A with nadir B directly beneath the platform. The microwave beam could be transmitted obliquely at right angles to the direction of flight illuminating a swath C in the ground plane. Range D refers to the across-track dimension perpendicular to the flight direction, while azimuth E refers to the along-track dimension parallel to the flight direction. The portion of the image swath closest to the nadir track of the radar platform is called the near range A while the portion of the swath farthest from the nadir is called the far range B (Figure 6.16). The incidence is the angle between the radar beam and ground surface A which increases, moving across the swath from near to far range. The look angle B is the angle at which the radar 'looks' at the surface. Depression angle γ, which is the complementary of the look angle, is the angle between the horizontal plane formed by azimuth and range, and the direction of the EM pulse to a point on the ground. It can vary between the near-range γ and far-range γ. Ground-range (GR) distance is the horizontal distance between the transmitter and the object. In order to obtain a mapping of the surface of interest, or an image, the returned signal is transformed by calculating slant range (SR), which is the distance between the transmitter and the object:

$$SR = c\Delta t/2$$
$$GR = \sqrt{SR^2 - H^2}$$
$$GR = H\sqrt{(1/\sin^2 \gamma) - 1}$$

(6.1.42)

where Δt is the time between transmission and echo reception. In order to distinguish between two targets, their echoes should not overlap. The pulse width τ represents the separation between both echoes. Therefore the ground range resolution between the two targets is

$$R_r = \tau \cdot c/(2\cos\gamma) = c/(2B_w\cos\gamma) \qquad (6.1.43)$$

where R_r is the ground range resolution, τ is the duration of transmission, B_w is the bandwidth and γ is the depression angle. This shows that the narrower the pulse width τ or the wider the bandwidth of the transmitted signal, the better the range resolution, but the signal-to-noise ratio decreases since there is less average energy. If pulses are too close together the system becomes range ambiguous. Also, as the depression angle increases the range resolution decreases, which means that the closer the target is to the nadir pointing of the sensor, the harder it is to resolve compared with distant objects.

Across-track resolution is dependent on the length of the pulse (P). Two distinct targets on the surface will be resolved (see Figure 6.17)

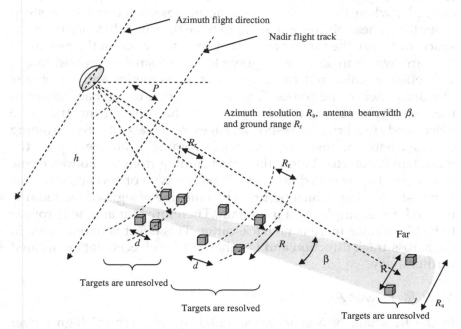

Figure 6.17 Target resolution[16]

in the range dimension if their separation is greater than half the pulse length.

Azimuth resolution depends on the position of the antenna and the beamwidth of the radar.

$$R_a = SR \cdot \beta = S \cdot \lambda/L = H\lambda/L \sin\gamma \qquad (6.1.44)$$

Where R_a is the azimuth resolution, SR is the slant range distance to the point of interest, λ is the wavelength of electromagnetic wave, L is the antenna length, H is the height of the sensor above the ground, γ is the depression angle and β is the beamwidth, which is a measure of the width of illumination pattern. For instance for a spacebased radar orbiting at 1000 km, antenna size is 1 m, incident angle is 30°, with frequency of 13 GHz and a bandwidth of 20 MHz, the $R_a = 26.6$ km and $R_r = 15$ m. For a given wavelength, the beamwidth can be adjusted by controlling the length of the antenna. Increasing antenna size will improve the azimuth resolution. Because of the limited mass and volume for space application, the increase of the antenna size could be done synthetically using SAR systems (see section 6.4).

Typical geometric distortion sources in radar imaging are fore-shortening, layover and shadowing. Foreshortening occurs, for example, when the sensor is looking at objects located on a steep mountain. These objects will seem to have the same distance from the spacecraft since the further away they are the higher on the mountain they are, which makes them appear to have a smaller ground range. This effect is enhanced for larger depression angles and for objects that are closer to the sensor. The accumulation of the backscatter of these multiple objects that is seen as if they come from the same place, leads to a brighter response. An extreme case of foreshortening occurs when, for instance, the object coming from the top of the mountain is detected before the object coming from the bottom of the mountain. This is called layover. The other type of geometric distortion is shadowing, which occurs when the line of sight of the radar is blocked, for example, by a mountain. Therefore that area will appear to be black since there is no backscatter. This effect is mainly seen in the across-track direction and is enhanced for objects that are located further away.

6.1.9 RF Power Amplifiers

In the RF chain of a spacebased radar system, typical high-power amplifiers used are traveling wave tube (TWT) amplifiers, Klystron

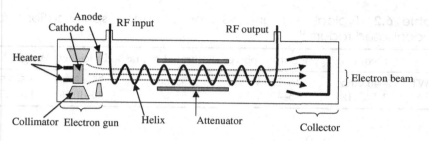

Figure 6.18 Typical TWT[10]

amplifiers, and solid-state power amplifiers (SSA). Each amplifier has characteristics that are better fit for one application than for another. The criteria in the amplifier selection are typically bandwidth, wavelength, mass, output power, linearity and cost.

The main advantage of a TWT amplifier over other tubes is the relatively broad frequency band of operation, with a typical bandwidth of 500 MHz plus high efficiency. Output power can range from a few watts to many kilowatts. One disadvantage is the higher cost and difficulty to operate and maintain, relative to Klystrons and SSAs. Figure 6.18 shows a typical TWT configuration. The electron gun, the slow-wave circuit and the collector constitute major components of the TWT. The electron gun produces a high-density electron beam via a heater that heats a cathode and a set of collimators and anodes to direct and guide the electron beam to the slow-wave circuit area. A control grid is often used to modulate the beam. Typically, a helix is used as a low-wave structure where the input of the RF wave is on the electron gun and the RF output is on the collector end. In this helix section, energy transfer occurs from the electron beam to the RF energy. This energy transfer or amplification occurs by slowing down the RF wave that travels along the helix at the velocity of light until it travels at the lower velocity of the electron beam. As electrons emerge from the slow-wave circuit, they are collected using a multi-stage design to dissipate their kinetic energy into heat that is conducted to the outside surface.

Solid-state amplifiers (SSA), are relatively cheap and reliable to use and, compared with TWTs and Klystrons, power is relatively limited (Table 6.2). They have a broad bandwidth, moderate gain, low noise and low efficiency. In phased array radars SSAs are used since RF transistors are easier to combine in large numbers. SSAs use either field effect transistors (FETs), or high electron mobility transistors

Table 6.2 Typical TWT and SSA performance specifications for spacebased radars[10]

	Frequency	RF output power	Saturated gain dB	Efficiency	Weight kg
TWT	2–40 GHz	10–250 W	50–60	55–65 %	2.5–3.5
SSA	2–5 GHz	5–40 W	55–65	30–40 %	1.5–2.5

(HEMTs). The gallium arsenide version of these transistors can achieve high-power input of up to 40 W in the S band with up to 40% efficiency. Yet the output power of the transistor is limited by the gate length, gate width and breakdown voltages.

Klystrons are less expensive, easier to set up, to operate and to maintain than TWTs. The bandwidths range from 40 to 80 MHz and are tunable over 500 MHz or more but much less efficient than TWT. A Klystron amplifier is basically an electron tube which uses velocity modulation to generate or amplify ultra-high frequencies. A cathode produces electrons which are then accelerated by a voltage pulse and focused into a beam using magnets. The electron beam moves past several cavities which are tuned to different frequencies to provide the required bandwidth. The input cavity modulates the velocity of the beam. The output cavity is then coupled to the transmission line. A collector then absorbs the high-energy electrons and dissipates the heat. The collector should be cooled.

6.2 Radar Imaging

Radar images are composed of pixels where each pixel represents an estimate of the radar backscatter signal for a given target. The backscattered signal depends on the interaction between the radar system and the target. The frequency, polarization, phase and viewing geometry of the radar system can react with variable characteristics of a target such as its geometry, electrical and composition properties. The brightness or the amplitude of an image is actually a combination of several variables. Empirical observation is one of the main aspects of interpreting reflected radar energy. The echo from a nadir or side-lit radar responds differently to the structure of the target, even though both rely on a backscattered signal, the first one is typically used for altimetry and the second one is used for most other sensing. The incident angle enhances the characteristics of the target, which make it more favourable for radar imaging.

The amplitude of the echo, which provides information about the target from a pixel in a radar image is recorded. The time required to travel to the target and back, is also recorded by the sensor. Commonly, a bright tone means a high backscatter and a dark tone corresponds to a low backscatter. The next pixel is recorded when the platform moves along the flight path or the radar beam is steered, to form a two-dimensional image.

For the geometric characteristics of the target, typically flat surfaces such as roads will backscatter much less towards the radar than rough surfaces. A surface is considered rough if the surface structure is comparable or higher than the wavelength. For a flat surface the reflected EM wave is directed away from the radar. Therefore flat surfaces will be darker than rough surfaces. In other words, the surface roughness of a feature controls how the microwave energy interacts with that surface, or target, and is generally the dominant factor in determining the tones seen on a radar image. For the orientation of the target, surfaces that are facing the radar would have a higher returned signal than the ones that are facing away from the radar line of sight. For mountains, foreshortening is a dominant effect with SAR systems. Mountains will have a higher backscatter from one side and no backscatter from the back slope which is located in the shadows.

Absorbed, transmitted and reflected EM waves depend on the electrical properties of the target and, more specifically, on the complex dielectric constant of the target since it is the indication of the reflectivity and conductivity of that target's material. This dielectric constant increases as the reflectivity and conductivity increases. Therefore any effect on this property should impact the response of the echoed radar signal and its appearance in the image. For instance, the dielectric constant of water is at least 10 times that of dry soil. Therefore, presence of moisture in soil should increase the dielectric constant which leads to higher radar reflectivity. Another property that leads to high radar reflectivity is metal. Metal objects such as bridges, pylons, silos and railroad tracks have high returns and appear on SAR images as bright spots. Any large-scale surface variation will also tend to appear bright on the side that faces the sensor and dim on the opposite side

Another type of interaction between the radar signal and the target is the polarization of the EM wave. The target can respond differently for an H or V polarization. It can also backscatter only one type of polarization and absorb or dissipate the other, which becomes the signature of that type of target. Therefore, radar signals can be tuned to transmit/receive in different modes of polarization. In the received mode, the signal can be filtered to just process one type of polarization. Typical combinations of transmit/receive polarized EM are HH for horizontal transmit and horizontal receive, VV for vertical transmit and vertical receive HV for horizontal transmit and vertical receive and VH for vertical transmit and horizontal receive.

6.3 Altimetry

Altimeters measure the round-trip travel time of microwave radar pulses to determine the distance from the spacecraft to a surface. The altitude is determined from the waveform of the returned signal. For ocean surfaces, the determination of this altitude leads to the analysis of ocean circulation and transports, hurricanes, forecasting and much more. This technique has been used for many years, which makes it a mature technique, and continuous improvements in accuracy have been made as models and computational techniques have improved. There have been major advances in sensor and algorithm processing over the years to bring the accuracy to within a few centimeters.

The absolute height of ocean surface topography above the reference ellipsoid (Figure 6.19) is called sea surface height SSH and is defined as

$$SSH = S - R = G + \eta$$
$$SSH_i = \delta\eta_i + SSH_{mean}$$

(6.3.1)

where the reference ellipsoid is the reference baseline for satellite altitude, G is the geoid height that is the gravity equipotential surface S is the spacecraft altitude from the reference ellipsoid, R is the range, η is the dynamic topography which is the height of ocean due to dynamics, and $\delta\eta$ is the residual topography which is the anomaly of SSH, subtracting out long-term mean SSH. G is still poorly known especially in areas where the topography of the surface varies greatly.

Altimeter measurements of sea surface topography are affected by many sources of disturbance such as ocean, atmospheric and orbit perturbation effects (Figure 6.19). These disturbances could be com-

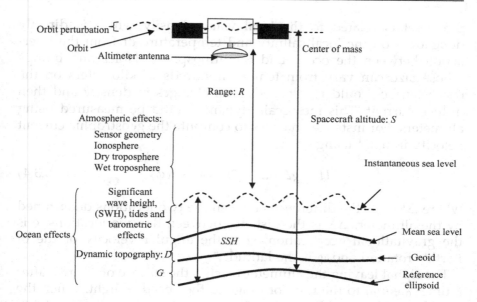

Figure 6.19 Typical disturbances source for an altimeter (not to scale)

pensated for empirically or with spatial averaging, modeling or on-board instruments. For a spacebased altimeter, the spacecraft flies the same orbit along the same ground tracks within a period of a few days, depending on the orbit parameters. Using this repeated track with a given period, one can therefore compute the mean height along each satellite ground track:

$$\langle SSH \rangle = G + \langle \eta \rangle \tag{6.3.2}$$

where the sign $\langle \ldots \rangle$ in this equation indicates a mean time average value over a great number of passes on the same track, considering G does not vary with time. This way the geoid anomaly is removed and η, which is the parameter of interest, is computed. By subtracting, at each point of the track, the instantaneous measurement from the time mean of the measurements at that point, one eliminates the geoids, and gets an anomaly of the dynamic topography, i.e the deviation from the mean with an excellent accuracy:

$$SSH - \langle SSH \rangle = \eta - \langle \eta \rangle = SLA. \tag{6.3.3}$$

Sea level is related to the local water density which is directly dependent on the local salinity and temperature or from the interaction between the ocean and the atmosphere. Holes and bumps whose size can vary from tens to hundreds of kilometers on the ocean surface could result from local changes in density and then induce current. This mesoscale dynamic could be measured using altimeters. For instance, in order to compute the geostrophic current velocity in ms^{-1} using

$$U = g\Delta \, SLA/(2\Omega\Delta X \cdot \sin\varphi) \tag{6.3.4}$$

where ΔSLA is the difference between the two locations determined by the altimeter, ΔX is the distance between the two locations, g is the gravitational acceleration, Ω is the angular velocity of the of Earth's rotation and φ is the latitude.[4]

The signal leaving the altimeter reaches the surface of interest after a time t_0 equal to the ratio of range to the speed of light. When the incident radar pulse first meets the surface it illuminates a point and a reflected signal begins to return to the satellite. This point expands to a disk as the wavefront continues to propagate (Figure 6.20). The maximum area of the surface of this disk is the pulse limited footprint which occurs when the back of the pulse has reached the ground before the leading edge extends to the antenna beam limits. The area A_f and the diameter D_f of this footprint for the ideal case is

$$A_f = \pi Rc\tau; D_f = 2\sqrt{Rc\tau} \tag{6.3.5}$$

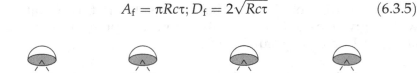

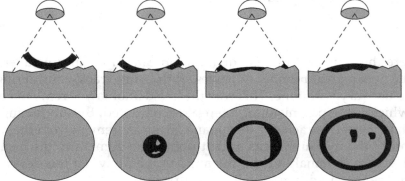

Figure 6.20 Waveform of a radar altimeter: illuminated area for a rough ocean surface

where τ is the pulsewidth and R is altitude of the spacecraft. Once this footprint is reached, an annulus is formed whose radius increases until it reaches the edge of the radar beam. For a given altimeter where $t = 3.03$ ns and $R \sim 800$ km, the area of the footprint A_f is about 2.3 km^2 and the diameter of the footprint D_f would be of the order 1.7 km. For rough ocean surfaces when waves are present, the first echo that reaches the spacecraft is the one returned from the scattering of the EM waves by the ocean wave return crest as illustrated in Figure 6.20. The rest of the EM wave is received at a later time. The echoed signal that reflects the ocean surface char- acteristics is contained in the broadened return time and the widened leading edge of the waveform. The pulse limited footprint diameter D_f is then

$$D_f = 2\sqrt{hc\tau'}$$

$$\tau' = \sqrt{\tau^2 + \left(4\sigma_s \ln 2/c\right)^2}$$

where σ_s is the RMS wave-height.[7]

6.4 Envisat: RA-2

The radar altimeter system RA-2 is an instrument that is part of Envisat (Figure 5.35). RA-2 is a nadir-looking pulse-limited radar altimeter with a main nominal frequency of 13.575 GHz (Ku Band) and a secondary channel at a nominal frequency of 3.2 GHz (S band) that is used to estimate the errors on range measurements induced by the ionosphere and the atmosphere. Geophysical corrections, that are either derived empirically, numerically or with other types of mea- surements, are used to minimize the delay caused by these errors. For errors induced by the heating and cooling cycles of the space- craft, the RA-2 periodically performs internal calibration measure- ments by coupling a calibration coupler into the receiver in the front RF electronics. This would indicate all errors and disturbances in the Tx/Rx path. This process does not include the antenna.

Table 6.3 summarizes the Tx/Rx signal characteristics, i.e. band- width, frequency and peak power for both the S and the Ku band. RA-2 transmits a 450 MHz chirp signal which is generated by a chirp generator. The three bandwidths used for the Ku band are 320, 80

Table 6.3 RA-2 key system parameters[6]

Parameter	Value
Orbit range	764–825 km
Operative frequency	13.575 GHz (Ku), 3.2 GHz (S)
Pulse width	20 μs
Ku Tx pulse bandwidth	320 - 80 - 20 MHz - CW
S Tx pulse bandwidth	160 MHz
Tx peak power	60 W (Ku), 60 W (S)
Pulse repetition interval	557 μs (Ku), 2228 μs (S)
Antenna gain	41.6 dBi (Ku), 29.2 dBi (S)
Antenna -3 dB beamwidth	1.35 deg (Ku), 5.5 deg (S)
IF center frequencies	1223 MHz/75 MHz
RF losses	1.8 dB (Ku), 1.7 dB (S)
Receiver maximum gain	107 dB
AGC dynamic range	60 dB
Receiver noise bandwidth	6.4 MHz
Receiver noise figure	3.0 dB (Ku), 2.5 dB (S)
Backscatter coefficient	−10 dB to +50 dB
Wave height	0.5 m to 20 m
Pulse repetition frequency	1795.33 Hz (for Ku-Band), 448.83 Hz (for S-Band)
IF bandwidth	6.4 MHz

and 20 MHz. For the S band, the bandwidth used is 160 MHz. This signal is up-converted to Ku and S bands. Both Ku and S signals are amplified and transmitted using the same 1.2 m diameter parabolic antenna (Figure 6.2). A 60 W peak power and 3.6% duty cycle TWT amplifier is used to transmit the Ku band signal. A 60 W peak power and 1% duty cycle SSA is used to transmit the S band signal. The antenna gain depends on the band used.

In the receiving mode, switched circulators are used for the Ku band, and PIN diodes for the S band are used to route the signal to a low-noise amplifier (LNA) and then to the de-ramp mixer (see Figure 6.21). It is critical that the switching mechanism used works properly in order to avoid any RF leaks from the high-power signal to the amplifier of the receiver that is geared for low-power signals. The bounced signal from the target of interest is a packet of multiple chirp signals reflected from each facet or section of the target. These echoed signals are then passed through an LNA, and then mixed with an intermediate frequency (IF) signal to mix down the time delays into constant frequency tones using a deramping mixer. The IF signal is actually a delayed replica of the transmitted chirp signal

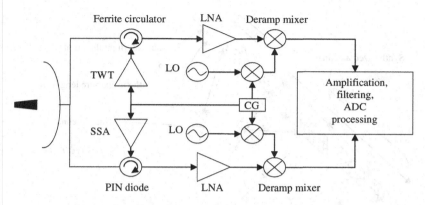

Figure 6.21   Block diagram of the RA-2[6]

that was up-converted using a local oscillator. The difference in frequency between two input signals is obtained using a deramping mixer. When the two inputs carry the same frequency change rate, the output frequencies are constant tones. These tones depend on the range of the target. The resulting signal is then filtered, amplified, down-converted to IF (6.4 MHz), sampled, digitized and processed.

6.5 Synthetic Aperture Radar

To acquire high azimuthal resolution, equation (6.1.44) shows that it could be done by increasing the size of the antenna. For spacebased radar, increasing the size of the antenna will increase mass, power and volume which is a very costly luxury. Another method would be to use the radar's echo Doppler history that results from the forward motion of the spacecraft to synthesize a long antenna equal to the distance the satellite traveled during the integration time. This method is called synthetic aperture radar (SAR). It uses a coherent microwave signal to synthesize a long antenna using a short one (Figure 6.22). The process in principle is simple but the implementation requires a complex integrated system, spacecraft orbital position and attitude, and antenna pointing control loops.

Figure 6.22 illustrates some of a SAR' parameters. It shows how an antenna can see the target as it travels; the overall observation can be seen synthetically as one antenna whose length is five times the physical dimension of the antenna. To understand the overall

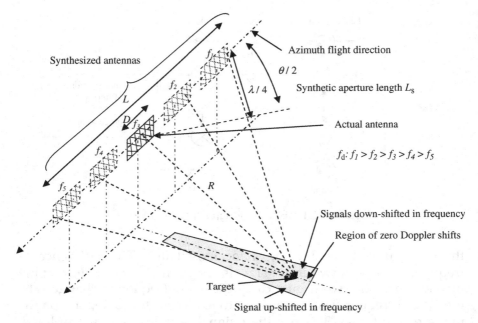

Figure 6.22 Illustration of a SAR system – the SAR length here is $L = 5D$

process of a SAR system, one can look at the response from one point target. The radiation pattern of a SAR target is

$$F_{\text{SAR}} = (\sin(u)/u)^2$$
$$u = \pi L \sin(\theta)/\lambda. \qquad (6.5.1)$$

The length L of the SAR system is bounded by the phase error when it reaches $\lambda/4$. Therefore,

$$L_{\max} \sin(\theta/2)/2 = \lambda/4 \Rightarrow L_{\max} \approx \sqrt{\lambda R}. \qquad (6.5.2)$$

Also, the beamwidth has to be wide enough to illuminate the target:

$$\left. \begin{array}{l} L \leq R\theta_{3\text{dB}} \\ F_{\text{SAR}}(\theta_{3\text{dB}}) = 0.5 \end{array} \right\} \Rightarrow u \approx 1.39 \Rightarrow \theta_{3\text{dB}} = \arcsin(0.443\lambda/L) \quad (6.5.3)$$

Therefore the azimuth resolution R_{aSAR} for small angles is:

$$R_{\text{aSAR}} = 0.5\sqrt{R\lambda}. \qquad (6.5.4)$$

This cross-range resolution did not account for the Doppler effect. The previous cross range is often referred to as the unfocused SAR system. Within the wide antenna beam, the returned signal from features that are ahead of the spacebased SAR will have higher frequencies because of the Doppler effect. The narrower areas behind the SAR system would have lower frequencies for the same reason. For the area located near the centerline beamwidth, there is no frequency shift. Processing the returned signal using the Doppler shift information leads to a narrower antenna beamwidth than using a synthetic antenna would. Looking at the cross range with the Doppler effect is really what SAR systems are about. Differentiating equation (6.1.41), the rate of change of the Doppler frequency is

$$\partial_t f_d = 2v^2 \sin(vt/R)/\lambda R. \tag{6.5.5}$$

The target stays within the beam for a duration ΔT. Referencing time to zero when $\theta = 0$,

$$\Delta f_d = 2v^2 \Delta T/\lambda R. \tag{6.5.6}$$

The azimuth resolution is then

$$\left.\begin{array}{l} R_{aSAR} = R_a \delta f/\Delta f_d \\ \delta f = 1/\Delta T \\ R_a = R\lambda/D = c\Delta T \end{array}\right\} \Rightarrow R_{aSAR} = D/2 \tag{6.5.7}$$

where R_a is the azimuth resolution of the real beam with aperture D. This equation shows that the azimuth resolution for this case is independent of the range.[10] The distance between pulses has to be less than half of the antenna size, and the spacecraft has to cover less than the azimuth length of the antenna between the emitted pulses in order to ensure a proper range resolution.

6.6 Envisat: ASAR

In addition to the RA-2 as presented in section 5.5, Envisat (Figure 5.35) carries the advanced synthetic aperture radar (ASAR) which is a high-resolution, wide-swath imaging radar instrument that can produce a two-dimensional view of a specific site of interest with high resolution. It is geared for land, sea, ice, ocean monitoring and surveillance. High-resolution is made possible because of its peak power capability, the pulse compression techniques used and pulse coherence. For the ASAR, transmitted and received beam patterns can be steered and shaped for different swaths by adjusting the gain and phase of the transmit/receive modules across the antenna. This allows, for example, multiple pass radar interferometry, aperture synthesis and different incident angles on the area of interest.

ASAR consists of a coherent signal, i.e. amplitude and phase are maintained over the collection interval. Each pulse starts with the same phase, with a dual polarization HH/VV or cross polarization HH and HV or VV and HH (where H stands for horizontal and V stands for vertical). This antenna is a 1.3 m × 10 m phased array (Figures 6.23 and 6.24) that is folded over in five panels that deploys after launch.[5] Each subarray is connected to a T/R module, and is a part of a two RF distribution corporate feed. The combination of the 24 ring slot per subarray provides a constant phase and amplitude illumination. The radiating characteristics of the subarrays are 20.5 dB in gain and a 1.5 dB in loss. The antenna has a high degree of stability since the temperature of each transmit/receive module is monitored and phase and amplitude are compensated for accordingly.

The received echo signal is routed to the RF distribution system using the RF front end and a corporate feed system which should be able to add the input signal coherently and noise incoherently.

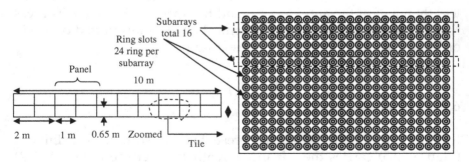

Figure 6.23 ASAR phased array antenna and its tiles

Folded panel that supports subarrys   Fully deployed array

Subarray of tiles

Figure 6.24   Envisat, ASAR flight model antenna (Courtesy of European
Space Agency ESA Remote Sensing Missions ©)

6.7 Interferometric SAR

In general, interferometry is directly related to resolution. Combining interferometry with SAR systems leads to the so called interfero-metric SAR system (InSAR). InSAR has proven to be valuable in a wide range of applications. It is especially useful in applications that relate to disaster management since, with this technique, displace-ments at the centimeter and sub-centimeter level resulting from seismic motion or other phenomena that could impact the surface of the Earth can be detected.

The implementation of this technique is to use two similar passes and interfere the images. A pass is a flyby of the instrument by the target. For classic interferometry (as illustrated in Figure 4.12) the base is the distance between the two receivers. However, for InSar (as illustrated in Figure 6.25) the base would be the distance between the two orbital locations where the images were taken at each pass. This method requires extensive processing which includes accurate orbit determination for each pass and repeated tracking. The duration

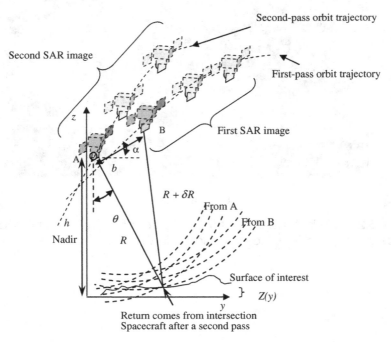

Figure 6.25 InSAR system with two passes

between passes can be in the order of hours or days depending on the orbit of the spacecraft. SAR systems can measure the phase of the echoed signal, which is used in the phase difference measurement for the InSAR. The phase difference between two measurements or images, which produces interferograms, results from the difference in starting points of the same emitted EM wave.

The radar phase is

$$\Delta\phi = \frac{2\pi}{\lambda}\delta R \Rightarrow \delta R = \frac{\lambda\Delta\phi}{2\pi} \qquad (6.7.1)$$

(for δR see Figure 6.25). Based on the geometry in Figure 6.25, the elevation of the point of interest is

$$z(y) = h - R\cos\theta \Rightarrow z(y) = h - \frac{1}{2}\left(\frac{(\lambda\Delta\varphi)^2 - (2\pi b)^2}{2\pi b\sin(\alpha - \theta) - (\lambda\Delta\varphi)}\right)\cos\theta.$$

$$(6.7.2)$$

The proper combination of two images taken of the same area but from two slightly different locations gives an interferogram. This interferogram is a map of the change in measured phase, and therefore the change in distance from the target to the spacecraft across the two SAR images. The phase difference in an interferogram has a 2π ambiguity, which results from the mathematical derivation of the phase because sine and cosine functions are periodic with period 2π. In order to correct for this ambiguity the phase has to be unwrapped, or referenced, which boils down to adding the appropriate number of 2π to the phase difference at each point in the image. The unwrapped image of a flat area would produce a constant level in the image.

Thus far, InSAR has been used to determine topographic height, for example, by looking at the interferogram that results from two images taken from two passes viewing the same target from virtually the same location. Using a third pass, a second interferogram can be derived from the second and third pass. Looking at the difference between these two interferograms, one can derive any displacement associated with tectonic movements, earthquakes and landslip for example. Displacement fringes are obtained from the subtraction of the two interferograms. These passes can be chosen before and after an event such as an earthquake. Subtraction of two interferograms is called differential interferometry. This method uses the same space-based instrumentation as a SAR system; however, interferograms are analyzed by product which requires a lot a computation with complex algorithms.

6.8 Ground Penetrating Radar (GPR)

Ground penetrating radar (GPR) is a technique that is used to unveil some subsurface features such as composition, structure and depth using low frequency EM waves. Subsurface features of Earth or Mars are composed of various materials that have different electric and magnetic permittivity and are possibly structured in layers. These layers will reflect EM waves accordingly. This technique has been proven and is widely used to find buried objects, or interfaces beneath the surface or embedded in opaque structure.

Figure 6.26 illustrates a typical implementation of this technique where low-frequency emitters transmit EM waves that penetrate the ground. The receiver collects the echoed signal from the various reflections of materials and structure. The reflection coefficient ρ of EM waves[3] at normal incidence is

$$\left. \begin{array}{l} \rho = |(Z_2 - Z_1)/(Z_2 + Z_1)| \\ Z_i = \sqrt{\mu_0/\varepsilon_0\varepsilon_{ri}} \end{array} \right\} \Rightarrow \rho = |(\sqrt{\varepsilon_{r1}} - \sqrt{\varepsilon_{r2}})/(\sqrt{\varepsilon_{r1}} + \sqrt{\varepsilon_{r2}})|,$$

$$(6.8.1)$$

where Z_i is the characteristic impedance of material i. Free space impedance is zero. ε_r is the relative electric permittivity and μ_r is the relative magnetic permittivity. This is generally a function of

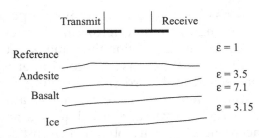

Figure 6.26 Example of a multilayer crust with a layer of ice, where ε here is the relative permittivity

frequency. The one-way attenuation (α_d) in dB for propagation through an unbounded dielectric material is

$$\alpha_d = 27.3\sqrt{\varepsilon_r}d\tan(\delta)/\lambda$$
$$\tan(\delta) = \varepsilon''/\varepsilon' \qquad\qquad (6.8.2)$$
$$\varepsilon = \varepsilon' - j\varepsilon''$$

where ε' is the electric permittivity, ε'' is the losses associated with both conductivity and frequency, $\tan(\delta)$ is the loss tangent and d is the distance.[3] For instance, searching for an ice packet embedded in the following crest structure one can use the various relative electric permittivities as per Table 6.4.

The reflection coefficient ρ and the transmission coefficient τ in the configuration of Figure 6.26 are: $\rho_{RA} = 0.3$, $\rho_{AB} = 0.18$, and $\rho_{BI} = 0.45$. Therefore $\tau_{RA} = 0.7$, $\tau_{AB} = 0.82$, and $\tau_{BI} = 0.55$, where the subscript R stands for reference, A stands for andesire, I stands for ice and B stands for basalt. Since the propagation velocity of the EM wave in a material i is

$$c_i = \frac{c}{\sqrt{\varepsilon_R}}$$

therefore the range resolution is

$$\delta R = \frac{c_i\beta}{2} = \frac{c\beta}{2\sqrt{\varepsilon_R}}$$

where β is the pulse width. For layers smaller than the wavelength, the scattering cross section could be approximated using the Rayleigh formula (section 4.2.2)

$$\sigma = \frac{128\pi^5}{3}\frac{a^6}{\lambda^4}.$$

Table 6.4 Relative permittivity and loss tangent for some materials[9]

	Crust material		Pore filling material	
	Andesite	Basalt	Ice	Water
Relative permittivity	3.5	7.1	3.15	88
Loss tangent	0.005	0.014	0.00022	0.0001

Therefore the radar range equation that includes loss due to attenuation (equation (6.8.2)) is

$$P_{\mathrm{R}} = P_{\mathrm{T}} + 2G_{\mathrm{ant}} + 10\log_{10}\left(\frac{\lambda^2}{(4\pi)^3}\right) + \sigma_{\mathrm{dB}} - \alpha_{\mathrm{d}}d$$

where d is the depth and L is the rest of the losses that are associated with the atmosphere and reflection from the surface.

6.8.1 MARSIS

Mars Global Surveyor and Mars Odyssey observations showed that there is the potential presence of a large amount of water below the Martian surface. The Mars Express orbiter, that would orbit Mars at 250 to 300 km, carried the Mars advanced radar for subsurface and ionosphere sounding (MARSIS, see Figure 6.27) which is an instrument that uses ground penetrating radar techniques to locate water or ice below the planet's surface. Figure 6.28 illustrates a typical interaction of MARSIS-like EM waves with a subsurface. The anticipated depth for observation is of the order of a few kilometers. The echoed signal would include information on the electrical properties

Figure 6.27 Illustration of the deployed antenna of MARSIS on the Mars Express (Courtesy of European Space Agency ESA)

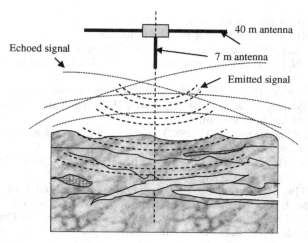

Figure 6.28 Example of MARSIS interaction with Martian subsurface[9]

of the reflected surfaces which is related to the composition of that surface, and the depth of that surface.

MARSIS is a 12 kg instrument that is composed of three subsystems. The primary antenna subsystem is a 40 m long high-efficiency dipole used to transmit and receive as illustrated in Figure 6.28. For transmit mode where RF signals from 1.3 MHz to 5 MHz are sent with a 1 MHz bandwidth, RF signals from 0.1 MHz to 5.4 MHZ are received with reduced efficiency. Its radiation gain peak is in the nadir direction. A second 7 m low-efficiency monopole antenna, which is physically orthogonal to the primary antenna and aligned with nadir, is used as a clutter cancellation antenna in the receive mode. The null of this antenna is in the nadir direction. The received signal is filtered, amplified, mixed and then digitized for processing, analysis and interpretation.[9]

Bibliography

1. Bahria, P. and Bahi, I. (1984) *Millimeter Wave Engineering and Applications*, John Wiley and Sons.
2. Barton, D. K. (1988) *Modern Radar System Analysis*, Artech House, Inc.
3. Daniels, D. J. (1996) *Surface Penetrating Radar*, The Institution of Electrical Engineers, London.
4. Elachi, C. (1987) *Introduction to the Physics and Techniques of Remote Sensing*, John Wiley & Sons.

5. ESA (2004) *Envisat ASAR Product Handbook.*
6. ESA (2004) *Envisat RA2-MWR Product Handbook.*
7. Levanon, N. (1988) *Radar Principles,* John Wiley and Sons.
8. Mohanty, N. (1991) *Space Communication and Nuclear Scintillation,* Van Nostrand Reinhold.
9. Picardi, G. *et al.* (1999) *The Mars Advanced Radar for Subsurface and Ionosphere Sounding MARSIS: Concept and Performace,* IEEE International Geoscience and Remote Sensing Symposium.
10. Reedy, E. K. and Eaves, J. L. (1987) *Principles of Modern Radars,* Van Nostrand Reinhold Company.
11. Rees, W. G. (1990) *Physical Principles of Remote Sensing,* Cambridge University Press.
12. Sivan, L. (1994) *Microwave Tube Transmitters,* Chapman and Hall.
13. Skolnik, M. I. (1980) *Introduction to Radar Systems,* McGraw-Hill.
14. Skolnik, M. I. (1990) *Radar Handbook,* McGraw-Hill Publishing Company.
15. Soumekh, M. (1999) *Synthetic Aperture Radar Signal Processing,* John Wiley and Sons.
16. Ulaby, F. T, Moore, R. K. and Fung, A. K. (1982) *Microwave Remote Sensing: Active and Passive,* Volume II, Addison-Wesley Publishing Company.
17. Ulaby, F. T, Moore R. K. and Fung, A. K. (1986) *Microwave Remote Sensing: Active and Passive,* Volume III, Addison-Wesley Publishing Company.

7

GPS

The attitude control system (ACS) of a spacecraft relies on a wide range of sensors for orbit and attitude determination. Based on requirements such as mass, power, cost and pointing, the appropriate set of sensors is used. Typical ACS sensors are: horizon sensors, sun sensors, star sensors, magnetometers, accelerometers, gyros and the global positioning system (GPS) receivers, as illustrated in Figure 7.1. These sensors are used to find position, orientation, velocity, angular acceleration, linear acceleration and angular velocity. GPS receivers are gaining ground in spacebased applications as an intrinsic part of the attitude control system of a spacecraft. They can be used as a clock source for some payload or flight software and hardware. GPS receivers are also used for some atmospheric measurements.

The following sections will cover GPS basic concepts, signal structure, GPS data and GPS receivers. This chapter will close with typical spacebased applications that use GPS systems.

7.1 GPS Overview

GPS is a US system; the equivalent Russian system is the global orbiting navigation satellite system (GLONASS). Table 7.1 compares the two systems. They have similar space segments but the signal structures are considerably different.

Spacecraft Sensors.   Mohamed M. Abid
© 2005 John Wiley & Sons, Ltd

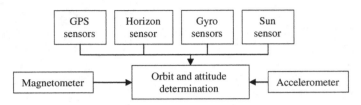

Figure 7.1 GPS concept

The GPS constellation consists of at least 24 US spacecraft that orbit the Earth in 12 h. To keep 24 working spacecraft accessible, new ones are launched to replace old ones. The orbit altitude is designed in a way that the spacecraft repeat the same track and configuration over any point approximately each 24 h. There are six high-altitude orbital planes with nominally four spacecraft in each, 60 degrees apart, and inclined at about 55 degrees with respect to the equatorial plane. This constellation allows any user on the Earth to view five to eight satellites at one time.

There are three generations of GPS spacecraft. Block 1 is the demonstration system that orbited at 63 degree inclination. Block 2 are the operational ones that orbit at 55 degrees and Block 2A are 'advanced' versions of Block 2. All these spacecraft are three-axis stabilized with nadir pointing using reaction wheels. They use S-band for control and telemetry communications and UHF cross-links between spacecraft. All spacecraft use L-band navigation signals at 1575.42 MHz (L1) and 1227.60 MHz (L2). For Block 1 the total power generated by the solar panel is 400 W. Block 2 power

Table 7.1 GPS/GLONASS comparison; n corresponds to the frequency channel

	GPS	GLONASS
Mean altitude	20 200 km	19 100 km
Orbital planes	6	3
Inclination	55 deg	65 deg
Signal separation	Code division multiple access spread spectrum	Frequency division multiple access
Carrier frequency	1575.42 MHz	$1602 + n*0.5625$ MHz
PRN code	32 possible CA codes	Identical maximal length for each n
Period	1023	511 chips
Rate	$1.023 \cdot 10^6$ chips/s	$511 \cdot 10^3$ chips/s
Data modulation	50 bits/s	100 bits/s

generation is about 710 W. Each spacecraft carries two rubidium and two caesium clocks. They also carry nuclear detonation detection sensors.

System performance, orbit prediction, orbit maintenance, constellation management, data error corrections, orbit ephemeris data, system status and GPS time are conducted by the master control station (MCS) which is located at Schriver AFB. Four other ground stations scattered around the world help in some of these tasks as well. This set of five stations constitutes the control segment.

In the user segment, GPS receivers are used for three-dimensional navigation, positioning, precise time determination, orbit determination, attitude determination, atmospheric characterizations and other research. GPS signals received by the user are converted into position, velocity and time estimates. A minimum of four satellites is required to compute the four dimensions of X, Y, Z (position) and time.

7.2 Concept

In order to gain a basic understanding of GPS receivers, a fog-horn example is used, as presented in Figure 7.2. Consider a fog-horn (A) that sounds daily at midnight using a timer. If this signal is heard at

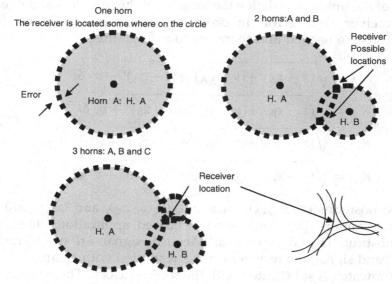

Figure 7.2 GPS Concept

5 s past midnight and the velocity of sound is $330\,\mathrm{ms^{-1}}$, one can deduce that the distance to the horn is 1650 m. With one horn, one can only determine the distance to the horn, the position could be anywhere on a circle with a radius of 1650 m and centered on the fog-horn (Figure 7.2). If the stopwatch that one was using to deduce the time is limited to 0.1 s accuracy, then the position is known within 33 m for that measurement. Thick lines in Figure 7.2 are used to illustrate this error.

Assume the receiver and the emitter have a synchronized clock, then the range R is

$$R = v_s(t_t - t_r) = v_s \cdot \Delta t \qquad (7.2.1)$$

where v_s is the sound velocity, t_t is the transmitted time and t_r is the received time. For the case where there is a clock error δt:

$$R_e = v_s(\Delta t + \delta t) = R + v_s \cdot \delta t. \qquad (7.2.2)$$

Using two more horns with different signals one can triangulate the position where all three horns were heard, as illustrated in Figure 7.2. Assuming that the clock timer for all three fog-horns are the same, and that the only uncertainty is in the receiver clock, then in the two-dimensional case, we have two unknowns: the distance and the receiver clock error. Therefore, four unknowns are present where three of the unknowns define the location of the receivers and one for the receiver clock error. In order to solve these unkowns, four equations are needed and therefore four fog-horns:

$$
\begin{aligned}
R_{eA} &= \sqrt{(x - x_A)^2 + (y - y_A)^2 + (z - z_A)^2} + v \cdot \delta t \\
R_{eB} &= \sqrt{(x - x_B)^2 + (y - y_B)^2 + (z - z_B)^2} + v \cdot \delta t \\
R_{eC} &= \sqrt{(x - x_C)^2 + (y - y_C)^2 + (z - z_C)^2} + v \cdot \delta t \\
R_{eD} &= \sqrt{(x - x_D)^2 + (y - y_D)^2 + (z - z_D)^2} + v \cdot \delta t
\end{aligned}
\qquad (7.2.3)
$$

where positions (x_A, y_A, z_A), (x_B, y_B, z_B), (x_C, y_C, z_C), and (x_D, y_D, z_D) are the transmitters' positions. For spacebased applications, fog-horns are substituted by GPS spacecraft. Acoustic signals are substituted by an L-band signal; the receiver ear is substituted with an antenna and the stopwatch is substituted with the receiver clock. The receiver is a GPS rather than a person.

7.3 GPS Signal

A GPS signal includes the GPS satellite identification, its position, the duration of the propagated signal and clock information. All this information is transferred through a robustly coded signal that is user specific, in that high precision is accessible to the military and reasonable resolution is accessible to other users.

7.3.1 Structure

The signal transmitted from the satellite is an EM wave. Two carrier signals in L-band, denoted L1 and L2, are generated by integer multiplication of f_o which is a fundamental frequency that is generated by an onboard oscillator. Carrier L3 is generated for military users only. GPS transmits at two microwave carrier frequencies. Table 7.2 summarizes the frequencies of the various codes embedded in the GPS signal.

The GPS signal is a phase modulated signal. The modulation form used is the binary phase shift keying (BPSK). GPS spacecraft transmit using code division multiple access spread spectrum (CDMA-SS). The emitted signal from a GPS spacecraft can be modeled by

$$
\begin{aligned}
s_{kL1}(t) &= \sqrt{2P_{YL1}}Y_k(t)N_k(t)\cos(2\pi f_{L1}t) \\
&\quad + \sqrt{2P_{CAL1}}CA_k(t)N_k(t)\sin(2\pi f_{L1}t) \\
s_{kL2}(t) &= \sqrt{2P_{L2}}Y_k(t)N_k(t)\cos(2\pi f_{L2}t) \\
Y_k(t) &= P_k(t)E_k(t)
\end{aligned}
\qquad (7.3.1)
$$

Table 7.2 GPS code frequencies

Fundamental frequency	f_o	10.23 MHz	29.30 m
Carrier L1	$2*77*f_o$	1575.42 MHz	19.0 cm
Carrier L2	$2*60*f_o$	1227.60 MHz	24.4 cm
CA code	$f_o/10$	1.023 MHz	293.05 m
P-codes	f_o	10.23 MHz	29.30 m
E-code	$f_o/20$	0.5115 MHz	586 m
Navigation message	$f_o/204\,600$	50 Hz	~ 6000 km

where P_{CAL1}, P_{YL1} and P_{L2} are the signal powers for signals carrying CA and Y codes on L1 and Y code on L2 respectively. The E code is used to encrypt the P-code to Y-code when anti-spoofing is implemented (this code is classified) and k is the number or the ID of the satellite.[2] The clear access or coarse acquisition (CA) code is a 1 MHz or 1023 bits (1 ms) repeating pseudo random noise (PRN) code which is a binary code defined as a function of time and generated using a tapped feedback shift register. For each GPS spacecraft a different CA code PRN is assigned. Precision or protected code (P-code), also characterized by a PRN sequence, has the fundamental GPS frequency f_o (10.23 MHz). This code modulates both the L1 and L2 carrier signals. The P-code is unclassified and defined in the GPS interface control document ICD-GPS-200. However, the encrypted P-code which is called Y-code and is referred to as the P(Y)-code is classified. The 0 or 1 in the signal is called chip. The encryption code has a 0.5115 MHz chipping rate which results in 20 P-code chips per encryption code chip. This insures that only authorized users can lock onto the Y-code, and have the capability to spoof it

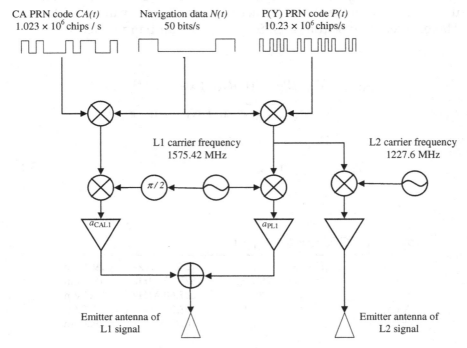

Figure 7.3 Modeled GPS signal diagram[5] (a is amplification gain)

Table 7.3 GPS signal parameters

Parameter	P-code	CA-code
Chipping rate	10.23×10^6	1.023×10^6
Chipping period	97.75 nsec	977.5 nsec
Range of one chirp	29.3 m	293 m
Code repeat interval	1 week	1 msec

by making a fake signal that would appears to be coming from a GPS satellite. Some correlation techniques could be used to lock onto the P(Y)-code.

Figure 7.3 shows schematically the construction of the modeled GPS signal using equations (7.3.1). The CA- and P-codes have characteristics that are summarized in Table 7.3. Because of the P-code repeat interval, typical GPS receivers lock first onto the CA-code since there is only 1 ms to search for accurate time information before locking onto the P-code.

7.3.2 GPS Data

In addition to the CA- or P(Y)-codes, the signal is also modulated with the 50 bits/s navigation message, which is composed of a set of 25 frames where each frame includes five subframes, as illustrated in Figure 7.4. Subframe one contains clock data parameters that describe the satellite clock and its relation to GPS time. This derives

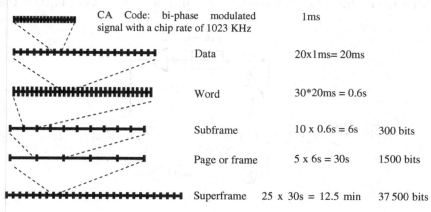

CA Code: bi-phase modulated signal with a chip rate of 1023 KHz		1ms	
Data		20x1ms= 20ms	
Word		30*20ms = 0.6s	
Subframe		10 x 0.6s = 6s	300 bits
Page or frame		5 x 6s = 30s	1500 bits
Superframe	25 x 30s = 12.5 min	37 500 bits	

Figure 7.4 GPS data format

clock corrections and range accuracy prediction. Subframes two and three include broadcast ephemeris data parameters, i.e. orbital data, of the transmitting satellite. The ephemeris parameters are used with an algorithm that computes the spacecraft position for any time within the period of the orbit. Subframes four and five are used to transmit different pages of system data such as the ionospheric model, satellite health, UTC parameters and military messages. The complete navigation message is composed of 125 subframes and sent over a 12.5 min period.[3]

The receiver extracts a great deal of information from this signal: L1 phase and pseudo range, L2 phase and pseudorange, CA on L1, Y-code typically for military or fake P-code for civilian, navigation message, Doppler observables, clock satellite status and spacecraft ephemeris such as mean motion difference, eccentricity, rate of inclination angle and argument of perigee.

7.4 GPS Receiver

The GPS receiver is composed of an antenna, front-end RF block, digital block and processing (Figure 7.5). Typically, a GPS receiver would take the input signal at the L1 or L2 band, filter it to isolate the GPS signal from the surrounding signals, down-convert the signal to intermediate frequency using a numerically controlled oscillator, and amplify the signal for digital sampling. Within an RF chain, filtering and amplification are repeated as needed.

The GPS spacecraft that originates the signal is identified by either CA codes or the Doppler shift. The number of receivers equals the

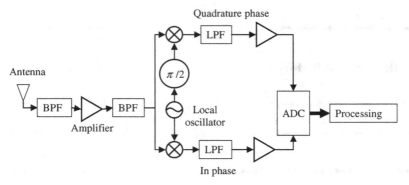

Figure 7.5 Typical GPS receiver[5]

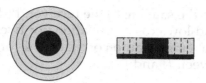

Figure 7.6 Choke antenna

maximum number of spacecraft that could be tracked simultaneously. GPS signals are transmitted with right-circular polarization
(RCP). RCP receiving antennas are preferred so that the antenna is
properly oriented to receive the signal. Two orthogonal dipoles could
be used. Other commonly used antennas are helical antennas, microstrips patches mounted closely to a flat plate and choke-ring antennas. The common goal of any GPS antenna is to minimize or control
multipath. Multipath occurs when the received signal comes from
the reflection of the GPS signal on the base where the antenna is
mounted or from adjacent objects. For spacebased GPS receivers
these objects could be solar arrays or surrounding instruments. To
overcome some of the multipath issues, the field of view of the
antenna should be obstacle free and a highly reflective plate should
be used with the antenna. Choke-rings have concentric rings on top
of a flat plate (Figure 7.6).

Multipath would slowly vary when the distance d between the
reflector and the antenna is of the order of the wavelength. Multipath
would vary rapidly and average to zero when d is much larger than
the wavelength.

7.5 GPS Signal Processing

GPS data could be extracted using various techniques. In the following section, an introduction to code phase and carrier phase techniques will be presented.

7.5.1 Code Phase Technique

Code phase is a processing technique that typically gathers data via a
CA code receiver. It uses pseudo random noise (PRN) codes to
identify which satellite the signal comes from and how long it

takes to travel from the satellite to the receiver. PRN codes have high auto-correlation and low cross-correlation, for two PRN sequences f_1 and f_2. The time averaged auto-correlation Ψ of f_1 and the cross-correlation Ψ_c between f_1 and f_2 are

$$\Psi(\tau) = \frac{1}{T} \int_0^T f_1(t)f_1(t - \tau)dt$$

$$\Psi_c(\tau) = \frac{1}{T} \int_0^T f_1(t)f_2(t - \tau)dt$$

(7.5.1)

where τ is the correlation width or the window size for acquisition. τ is inversely proportional to the bandwidth of the signal. The bandwidth of CA code is 1 MHz. Therefore, τ is 1 µs. The $P(y)$ side bandwidth is 10 MHz. Therefore, τ is 0.1 µs. Typically, the receiver generates CA codes based on its internal clock. The receiver-generated signals are correlated to the received signals until a match is obtained. This match identifies the emitter GPS satellite since each GPS satellite has its own PRN code. The time shift of the signal peak is then used to determine the distance between the receiver and the GPS satellite.

Figure 7.7 shows a typical correlation of an incoming PRN code with the one generated by the receiver. The correlation would determine the originating GPS spacecraft, and the offset of the correlation between the matched PRN gives the pseudorange.[2] The time difference between the transmitted signal based on the satellite clock and the received signal based on the GPS receiver clock leads to the range if both clocks were synchronized. Since both clocks drift, the time difference leads to the so-called pseudorange:

$$P_{rs} = (t_r - t_s) \cdot c$$

(7.5.2)

where P_{rs} is the pseudorange between receiver and satellite, t_r is the receiver clock time, t_s is the satellite transmit time and c is the speed of light. This expression can be related to the true range by introducing corrections to the clock times.

$$t_r = \tau_r + \Delta t_r \quad t_s = \tau_s + \Delta t_s$$

(7.5.3)

τ_r and t_s are true times, and Δt_r and Δt_s are clock corrections. Since the propagation velocity c depends on the media where the

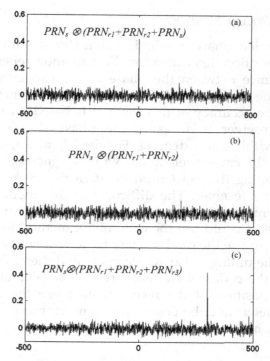

Figure 7.7 (a) Correlation between the received signal and the sum of two generated signals plus the incoming signal; (b) uncorrelated incoming signal with the sum of two PRN signals generated by the receiver; (c) correlation between the incoming PRN signal with a matched PRN generated by the receiver

signal propagates, substituting into the equation of the pseudorange yields:

$$P_{rs} = [(\tau_r - \tau_s) + (\Delta t_r - \Delta t_s)] \cdot c$$
$$P_{rs} = \rho_{rs} + (\Delta t_r - \Delta t_s) \cdot c + A_{rs}$$

$$(7.5.4)$$

where A_{rs} is the delay and ρ_{rs} is true range. Looking at the two-dimension pseudorange case, the navigation equation is

$$\rho_A = R_{eA} - v.\delta t = \sqrt{(x - x_A)^2 + (y - y_A)^2}$$
$$\rho_B = R_{eB} - v.\delta t = \sqrt{(x - x_B)^2 + (y - y_B)^2}.$$

$$(7.5.5)$$

7.5.2 Carrier Phase Method

Using the carrier phase method, sub-meter accuracy could be achieved for position determination. This method relies on measuring the difference between the phase of the carrier signal that is received by the receiver and the carrier phase generated by the transmitter. The accuracy of this method is based on the capability of the GPS receiver to measure the phase shift of the carrier frequency to within a few degrees. For the L1 carrier, a few degrees are a fraction 19 cm wavelength. With this method, the receiver would be counting the exact number of carrier cycles between the satellite and the receiver. The difficulty in this method is that all cycles are identical. To approximate the number of these cycles initially, receivers could use the code phase method to approximate this number then use the carrier phase to determine the actual cycles that identify the timing pulse. Receivers would therefore use the fast processing of the codes phase measurement to estimate with a low accuracy the position of the receiver, then use the carrier phase technique to determine the position with much more accuracy.

The carrier phase measurement is

$$\varphi_{rs}(t) = \frac{1}{\lambda}\left(R_{rs} + A_\varphi + c(\delta t_r - \delta t_s)\right) + N + \varepsilon_\varphi \qquad (7.5.6)$$

where R_{rs} is the range from the receiver to the GPS satellite, $A\varphi$ is the intervening media delay, N is the integer ambiguity and ε_φ is the carrier phase noise.[4] This technique requires a continuous direct sight of at least four GPS spacecraft in order to ensure a constant lock for position determination. This would ensure a constant integer ambiguity. The integer ambiguity could therefore be eliminated by making two phase measurements at different times, then taking the difference:

$$\varphi_{rs}(t_1) - \varphi_{rs}(t_2) = \frac{1}{\lambda}(R_{rs}(t_1) - R_{rs}(t_2)) + \Delta\varepsilon_\varphi. \qquad (7.5.7)$$

7.5.3 GPS Error Sources

GPS errors are a combination of noise, bias and blunders. Table 7.4 illustrates some of the noise sources and estimates of the error to GPS measurements. Errors do not all apply to spacebased receivers.

Table 7.4 Error source/cause/implication[4]

	Error source	Cause	Error
Blunders	Control segment mistakes	Computer or human error	1 m to hundreds of km
	User mistakes	Incorrect geodetic datum selection	1 m to hundreds of m
	Receiver errors	Software or hardware failures	Errors of any size
	Noise and bias errors		15 m for each satellite used in the position solution.
Bias errors	Selective availability	Intentional degradation of the GPS signal by the DoD	Various
	Satellite clock errors	Uncorrected by control segment	1 m
	Ephemeris data		1 m
	Tropospheric delay	Changes in temperature, pressure, humidity	1 m
	Ionosphere delays		10 m
	Multipath		~0.5 m

Selective availability is when the DoD intentionally dithers the GPS signal. This could be done by adding, for example, an unknown frequency to the GPS's fundamental frequency which degrades position in real time. Only specific users would be able to decipher the dithering to extract the exact information.

7.5.4 GPS Clock

GPS satellites use atomic clocks which are very stable with a high precision that relay on electrical oscillators regulated by the natural vibration frequencies of atomic systems such as caesium, hydrogen or rubidium atoms. Since atoms can have only discrete energy levels, the change or transition in energy between two levels corresponds to a frequency. An atomic clock locks an oscillator to one of these frequencies.

Caesium 133 is commonly used for atomic clocks. Atoms boiled off from heated caesium pass through a magnetic field that selects atoms of the appropriate energy state, and then they pass through a microwave field situated within a tube (Figure 7.8). When a caesium atom receives microwave energy at exactly the transition frequency

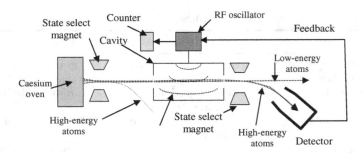

Figure 7.8 Typical caesium atomic clock

f_t, 9 192 631 770 Hz, it changes its energy state. The source of the microwave energy that is controlled by a quartz crystal oscillator would sweep a narrow range that includes this frequency. A magnetic field is then used to steer atoms that have changed their energy state toward a detector. The output of the detector is proportional to the number of incident caesium atoms. The maximum number occurs when the frequency of the microwave is equal to the excitation of a caesium atom. This maximum is used to lock the frequency of the crystal oscillator that controls the microwave source, which is divided by f_t to give the required one pulse per second.

Based on special relativity, GPS satellite clocks run slowly relative to Earth clocks since GPS spacecraft are moving at high speed relative to Earth. This motion slows clocks by a factor that is typically $< 10^{-8}$

$$\left(1 - \frac{1}{2}\frac{v^2}{c^2}\right) \tag{7.5.8}$$

where v is the speed of the spacecraft relative to Earth and c is the speed of light. Based on general relativity, GPS clocks run fast relative to Earth clocks since spacecraft are located in a weaker gravitational field relative to Earth. Gravitational potential slows clocks by a factor that is typically $\leq 10^{-8}$

$$\left(1 - \frac{GM}{rc^2}\right) = \left(1 + \frac{\phi}{c^2}\right). \tag{7.5.9}$$

All things considered, the relativistic net effect leads to the fact that orbiting clocks run faster than the clock on the ground by about 440 parts in 1012.

GPS clocks are monitored by the control segment. It is measured in weeks and seconds from 24:00:00, January 5, 1980 and is steered to within 1 μs of the universal time coordinate (UTC). UTC is computed from GPS time using the UTC correction parameters sent as part of the navigation data bits. Currently GPS time is ahead of UTC by 13 s. Part of the navigation data is clock correction information (a_0, a_1, a_2):

$$t_c = a_0 + a_1(t - t_{oe}) + a_2(t - t_{oe})^2 \qquad (7.5.10)$$

where t_c is the satellite clock correction at time t, t_{oe} is the time of ephemeris, a_0 is the clock bias, a_1 is the clock drift and a_2 is the clock drift rate.

7.6 GPS Coverage

The GPS system was originally designed for terrestrial and low-altitude applications. The transmission of a GPS signal is pointed towards the center of the Earth. The main beam at L1 is 42.6° wide, and at L2 is 46.8° wide, as illustrated in Figure 7.9. Figure 7.9 shows that receivers could be placed anywhere within the beam. This includes LEO, GEO and any intermediate orbits. Using side lobes of GPS emitters, this coverage can be extended. The duration of a GPS spacecraft that a spacebased receiver would see depends on the receiver orbit. Typically for a GEO receiver, GPS spacecraft would be in the field of view for ~ 15 min before it becomes shadowed by

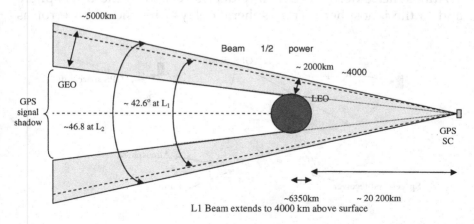

Figure 7.9 Main-lobe coverage of a GPS signal

Earth. For an LEO receiver the duration is in the order of 45 min. On the ground, the duration is in the order of 5 h. The number of GPS spacecraft in the field of view of a receiver depends on the orbit of this receiver. For instance, for a GEO receiver, typically four GPS spacecraft could be seen. Additionally, using GPS side-lobes signals, up to 10 spacecraft could be seen.

Some of the difficulties that spacebased receivers face are, for instance, high velocity of the emitter and receiver which leads to large Doppler offset and signal dynamics. It can be as large as 55 kHz for LEO compared with the 5 kHz for ground application. Carrier and code tracking loops have to maintain lock under high dynamics.

7.7 GPS for Atmospheric Measurements

One GPS error source is attributed to atmospheric delays in ground-based GPS signals. For GPS receivers used on spacecraft for attitude or orbit determination, this issue is irrelevant because the GPS signal used does not pass through the atmosphere. However, spacebased GPS receivers could be used to analyze the atmosphere for climate change and atmosphere characterization. This technique is called GPS radio occulation technique. Figure 7.10 illustrates a typical configuration of a spacebased receiver that collects GPS signals from GPS spacecraft when their signal bends as it travels through the atmosphere. Figure 7.11 shows an approximate thickness of Earth's atmospheric structure.

Atmospheric delays are mainly delays related to the troposphere and to the ionosphere. Tropospheric delays can lead to an error as

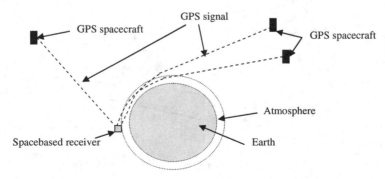

Figure 7.10   Typical spacebased GPS receiver for atmospheric study[2]

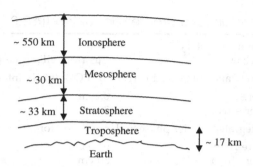

Figure 7.11 Earth's atmospheric structure and GPS recievers

high as 30 m and they are related to the changes in temperature, pressure and humidity. Knowing these delays, instant atmospheric tropospheric status could be derived by comparing these delays with the actual position and velocity of the receiver. The emitted signal is banded as it travels through the atmosphere (Figure 7.10). The error introduced by the ionosphere is in the 10–100 m range. This wide range depends on the time of the day, the season, the latitude and the path length that the signal travels through the ionosphere before reaching the receiver. A portion of this delay could be approximated by

$$\Delta t = \frac{40.3}{f^2} \cdot TEC \qquad (7.7.1)$$

where f is the carrier frequency of the GPS signal, and TEC is the influence of the total electron content which is the number of electrons in a 1 m^2 cross-sectional tube along the path of transmission through the ionosphere.[2]

The Challenging Minisatellite Payload (CHAMP) satellite was launched into an almost circular, near polar ($i = 87°$) orbit with an initial altitude of 454 km. It carries the turbo rogue space receiver (TRSR-2) GPS receiver. This receiver has a triple mode function. The first mode is the tracking mode used for orbit determination. The second mode is the occulation mode used for atmospheric limb sounding. The third mode is the altimetry mode used to measure the specular reflections of GPS signals from ocean surfaces.

Table 7.5   Performance characteristics of the TRSR-2

Computed position in telemetry	< 60 m
Time calibration accuracy	< 1 μs from GPS time (resolution 0.1 ns)
Dual-frequency range and integrated carrier for POD at a 1s interval:	
Phase (ionosphere-free)	< 0.2 cm
Range (ionosphere-free)	< 30 cm
Dual-frequency integrated carrier phase and amplitude for atmospheric occultation:	
Phase (ionosphere-free)	< 0.05 cm
Range (ionosphere-free)	< 50 cm
Limb-sounding observables (prior to atmospheric de-focusing):	
L1 carrier phase	< 0.05 cm (1 s)
L2 carrier phase	< 0.15 cm (1 s)

The TSRS-2 receiver can acquire up to 12 satellites at the same time. A navigation solution is obtained by using at least four GPS spacecraft. Table 7.5 illustrates some of the performance characteristics of the TRSR-2. For the tracking mode, TRSR-2 uses a zenith mounted antenna on a choke ring and a second antenna on the aft panel. For the occultation mode the high-gain helix antenna for occultation measurements is mounted on the aft panel. For the altimetry mode a high-gain nadir-viewing helix antenna is used. They are all illustrated in Figures 7.12 and 7.13.

7.8 Docking/Rendezvous

GPS can be used in docking/rendezvous maneuvers. Docking is often composed of three phases. The first phase is the homing phase. The separation between the target and the chaser is less than 50 km. The second phase is the intermediate station-keeping phase where the distance is within the 1 km range. A docking maneuver is concluded with the terminal phase that marks the end of the docking. In these three steps GPS could be used for various accuracy requirements. This would decide the type of GPS acquisition mode. In the carrier phase differential GPS or other can be used.

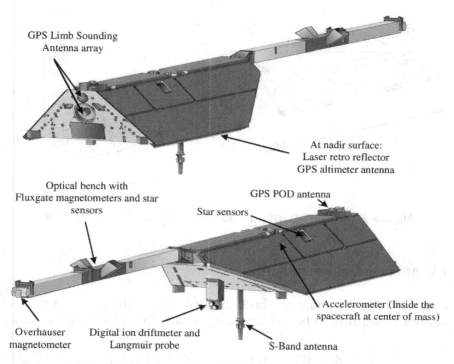

Figure 7.12 CHAMP spacecraft (Courtesy of © GFZ-Potsdam, Germany)

Figure 7.13 TRSR-2 antennas (Courtesy of © GFZ-Potsdam, Germany)

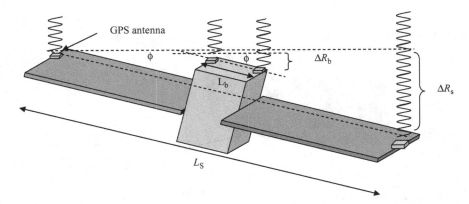

Figure 7.14 GPS configuration for attitude determination

7.9 Attitude Determination

Using the carrier phase method, multiple GPS antennas on a vehicle allow the attitude to be determined, as illustrated in Figure 7.14, so

$$\phi = \sin^{-1}(\Delta R_b / L_b) = \sin^{-1}(\Delta R_s / L_s) \tag{7.9.1}$$

where ϕ is the roll angle, L_b is the baseline between the two antennas located on the bus, L_s is the distance of the antennas located on the end of the solar panel and ΔR is the range difference between two antennas. Note that adding antennas on the solar panels is less trivial than adding them on the bus, but the roll-angle resolution is much easier to measure. Increasing the distance between the receivers has the advantage of obtaining a higher precision on ϕ. Unfortunately, this may lead to a more difficult tracking of the carrier phase measurements. Using equation (7.5.6), the carrier phase of each receiver from satellite s is

$$\varphi_{rs}(t) = \frac{1}{\lambda} \left(R_{rs} + A_\varphi + c(\delta t_r - \delta t_s) \right) + N + \varepsilon_\varphi.$$

Therefore the phase difference is

$$\Delta \varphi_s(t) = \varphi_{1s} - \varphi_{2s} = \frac{1}{\lambda} \left(R_{1s} - R_{2s} + c(\delta t_{1s} - \delta t_{2s}) \right) + N_{1s} - N_{2s} + \varepsilon.$$

For a baseline less than the L1 band wavelength 19 cm, N can only take the values -1, 0 or 1. Using the same clock for both receivers, this equation becomes

$$\Delta\varphi_s(t) = \frac{1}{\lambda}\Delta R_s + \Delta N_s + \varepsilon = \frac{1}{\lambda}L \cdot \sin(\phi) + \Delta N_s + \varepsilon.$$

The difference ΔR_s does depend on the GPS satellite in view and on the baseline between the two receivers.[4] Therefore,

$$\Delta\varphi_s(t) = \frac{1}{\lambda}\vec{e}_s \cdot \vec{L} + \Delta N_s + \varepsilon$$

where e is the unitvector pointing from the receiver to the GPS spacecraft s. Now looking at other satellites this equation becomes

$$\begin{pmatrix} \Delta\varphi_1 \\ \vdots \\ \Delta\varphi_n \end{pmatrix} = \frac{1}{\lambda}\begin{pmatrix} \vec{e}_1 \\ \vdots \\ \vec{e}_n \end{pmatrix} \cdot \vec{L} + \begin{pmatrix} \Delta N_1 \\ \vdots \\ \Delta N_n \end{pmatrix} + \varepsilon.$$

A Kalman filter could be used to resolve the ambiguity and noise terms. Kinematic modelings are used to describe the attitude and change in attitude of the set of receivers.

This approach could be extended for large spacecraft, such as the space station, to monitor structure deformation for example. This is mainly due to the high resolution of the carrier phase method.

7.10 AMSAT-OSCAR 40 (AO-40)

Earth orbit GPS experiment, AMSAT OSCAR 40 (AO-40) is a NASA GSFC experiment. The purpose of the GPS for this mission is to map a GPS constellation antenna pattern in space within and above the constellation to extend the use of GPS signal from Earth applications to spacebased applications for orbit and attitude determination. The orbit of the AO-40 is a high-Earth orbit (HEO) with the following orbit parameters: $a = 36\,286.409$ km, $e = 0.796838$, $i = 6.04$ deg, $\Omega = 151.818$ deg, $\omega = 335.84$ deg (see section 1.9). The nature of the orbit allows both LEO-like orbit receivers and GEO like orbit receivers to be covered. Doppler shift, however, is different for both cases. GPS hardware included four patch antennas on the perigee

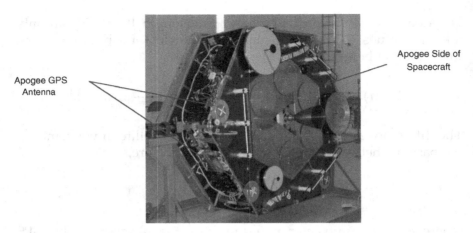

Apogee GPS
Antenna

Apogee Side of
Spacecraft

Figure 7.15 AMSAT Phase 3-D Spacecraft (Courtesy of AMSAT-DL–
picture by W. Gladisch ©)

side of the spacecraft and four high-gain antennas on the apogee side
of the spacecraft (Figure 7.15).

Bibliography

1. Farrell, J. A. and Barth, M. (1999) *The Global Positioning System and Inertial Navigation*, McGraw-Hill.
2. Hofmann-Wellenhof, B., Lichtenegge H. and Collins, J. (2001) *GPS Theory and Practice*, 5th edn, Springer, Wien and New York.
3. ICD-GPS-200C (1993) *Navstar GPS Space Segment/Navigation User Interfaces*.
4. Misra, P. and Enge, P. (2001) *Global Positioning System*, Ganga-Jamuna Press.
5. Tsui, J. B. Y (2000) *Fundamentals of Global Positioning System Receivers*, John Wiley and Sons Inc.

Index

Spacecraft Sensors. Mohamed M. Abid
© 2005 John Wiley & Sons, Ltd